我思

敢于运用你的理智

西方哲学经典影印

01. 第尔斯（Diels）、克兰茨（Kranz）：前苏格拉底哲学家残篇（希德）
02. 弗里曼（Freeman）英译：前苏格拉底哲学家残篇
03. 柏奈特（Burnet）：早期希腊哲学（英文）
04. 策勒（Zeller）：古希腊哲学史纲（德文）
05. 柏拉图：游叙弗伦 申辩 克力同 斐多（希英），福勒（Fowler）英译
06. 柏拉图：理想国（希英），肖里（Shorey）英译
07. 亚里士多德：形而上学，罗斯（Ross）英译
08. 亚里士多德：尼各马可伦理学，罗斯（Ross）英译
09. 笛卡尔：第一哲学沉思集（法文），Adam et Tannery 编
10. 康德：纯粹理性批判（德文迈纳版），Schmidt 编
11. 康德：实践理性批判（德文迈纳版），Vorländer 编
12. 康德：判断力批判（德文迈纳版），Vorländer 编
13. 黑格尔：精神现象学（德文迈纳版），Hoffmeister 编
14. 黑格尔：哲学全书纲要（德文迈纳版），Lasson 编
15. 康德：纯粹理性批判，斯密（Smith）英译
16. 弗雷格：算术基础（德英），奥斯汀（Austin）英译
17. 罗素：数理哲学导论（英文）
18. 维特根斯坦：逻辑哲学论（德英），奥格登（Ogden）英译
19. 胡塞尔：纯粹现象学通论（德文1922年版）
20. 罗素：西方哲学史（英文）

西方科学经典影印

01. 欧几里得：几何原本，希思（Heath）英译
02. 阿基米德全集，希思（Heath）英译
03. 阿波罗尼奥斯：圆锥曲线论*，希思（Heath）英译
04. 牛顿：自然哲学的数学原理，莫特（Motte）、卡加里（Cajori）英译
05. 爱因斯坦：狭义与广义相对论浅说（德英），罗森（Lawson）英译
06. 希尔伯特：几何基础 数学问题（德英）*，汤森德（Townsend）、纽苏（Newson）英译
07. 克莱因（Klein）：高观点下的初等数学：算术 代数 分析 几何，赫德里克（Hedrick）、诺布尔（Noble）英译

古典语言丛书

01. 麦克唐奈（Macdonell）：学生梵语语法
02. 迪罗塞乐（Duroiselle）：实用巴利语语法
03. 艾伦（Allen）、格里诺（Greenough）：拉丁语语法新编
04. 威廉斯（Williams）：梵英大词典
05. 刘易斯（Lewis）、肖特（Short）：拉英大词典

西方科学经典影印

The Works of Archimedes

阿基米德全集

（英文）

〔古希腊〕阿基米德（Archimedes）著
〔英〕希思（Thomas Heath）译

长江出版传媒 | 崇文书局

图书在版编目（CIP）数据

阿基米德全集：英文 /（古希腊）阿基米德
（Archimedes）著；（英）希思（Thomas Heath）译. --
武汉：崇文书局，2022.8
（西方科学经典影印）
ISBN 978-7-5403-6764-0

Ⅰ. ①阿… Ⅱ. ①阿… ②希… Ⅲ. ①古典数学-古希腊-文集-英文 Ⅳ. ① O115.45-53

中国版本图书馆CIP数据核字（2022）第 096231 号

阿基米德的著作，迄今发现的共有以下十一种：《论球和圆柱》《圆的度量》《论拟圆锥与旋转椭体》《论螺线》《论平面图形的平衡》《数沙者》《抛物线弓形求积》《论浮体》《引理汇编》《牛群问题》《方法》。
英国数学史家希思（Thomas Heath，1861—1940）在 1897 年翻译出版了前 10 种，1912 年翻译出版了第 11 种，都是由剑桥大学出版社出版。今即将此二书合刊影印，而 1897 年版的书前有希思撰写的长篇"导读"，影印时不收。

阿基米德全集（英文）

出 版 人	韩　敏
出　　品	崇文书局人文学术编辑部・我思
策 划 人	梅文辉（mwh902@163.com）
责任编辑	梅文辉
装帧设计	甘淑媛
出版发行	长江出版传媒　崇 文 书 局
地　　址	武汉市雄楚大街 268 号 C 座 11 层
电　　话	（027）87677133　邮政编码　430070
印　　刷	湖北新华印务有限公司
开　　本	880mm×1230mm　1/32
印　　张	12.25
字　　数	140 千
版　　次	2022 年 8 月第 1 版
印　　次	2022 年 8 月第 1 次印刷
定　　价	220.00 元

（读者服务电话：027—87679738）

本作品之出版权（含电子版权）、发行权、改编权、翻译权等著作权以及本作品装帧设计的著作权均受我国著作权法及有关国际版权公约保护。任何非经我社许可的仿制、改编、转载、印刷、销售、传播之行为，我社将追究其法律责任。

THE WORKS

OF

ARCHIMEDES

EDITED IN MODERN NOTATION

WITH INTRODUCTORY CHAPTERS

BY

T. L. HEATH, Sc.D.,

SOMETIME FELLOW OF TRINITY COLLEGE, CAMBRIDGE.

CAMBRIDGE:
AT THE UNIVERSITY PRESS.
1897

[*All Rights reserved.*]

This new DOVER EDITION *is an unabridged reissue of the* HEATH EDITION *of 1897 and includes the Supplement of 1912*

By special arrangement with The Cambridge University Press

PREFACE.

THIS book is intended to form a companion volume to my edition of the treatise of Apollonius on Conic Sections lately published. If it was worth while to attempt to make the work of "the great geometer" accessible to the mathematician of to-day who might not be able, in consequence of its length and of its form, either to read it in the original Greek or in a Latin translation, or, having read it, to master it and grasp the whole scheme of the treatise, I feel that I owe even less of an apology for offering to the public a reproduction, on the same lines, of the extant works of perhaps the greatest mathematical genius that the world has ever seen.

Michel Chasles has drawn an instructive distinction between the predominant features of the geometry of Archimedes and of the geometry which we find so highly developed in Apollonius. Their works may be regarded, says Chasles, as the origin and basis of two great inquiries which seem to share between them the domain of geometry. Apollonius is concerned with the *Geometry of Forms and Situations*, while in Archimedes we find the *Geometry of Measurements* dealing with the quadrature of curvilinear plane figures and with the quadrature and cubature of curved surfaces, investigations which "gave birth to the calculus of the infinite conceived and brought to perfection successively by Kepler, Cavalieri, Fermat, Leibniz, and Newton." But whether Archimedes is viewed as the man who, with the limited means at his disposal, nevertheless succeeded in performing what are really *integrations* for the purpose of finding the area of a parabolic segment and a spiral, the surface and volume of a sphere and a segment of a sphere, and the volume of any segments of the solids of revolution of the second degree, whether he is seen finding the centre of gravity of a parabolic segment, calculating arithmetical approximations to the value of π, inventing a system for expressing in words any number up to that which we should write down with 1 followed by 80,000 billion ciphers, or inventing the whole science of hydrostatics and at the same time carrying it so far as to give a most complete investigation of the positions of rest and stability of a right segment of a paraboloid of revolution floating in a fluid, the intelligent reader cannot fail to be struck by the remarkable range of subjects and the mastery of treatment. And if these are such as to create genuine enthusiasm in the student of Archimedes, the style and method are no less irresistibly attractive. One feature which will probably most impress the mathematician accustomed to the rapidity and directness secured by the generality of modern methods is the *deliberation* with which Archimedes approaches the solution of any one of his main problems. Yet this very characteristic, with its incidental effects, is calculated to excite the more admiration because the method suggests the tactics of some great strategist who foresees everything, eliminates everything not immediately conducive to the execution of his plan, masters every position in its order, and then suddenly (when the very elaboration of the scheme has almost obscured, in the mind of the spectator, its ultimate object) strikes the final blow. Thus we read in Archimedes proposition after proposition the bearing of which is not immediately obvious but which we find infallibly used later on; and we are led on by such easy stages that the difficulty of the original problem, as presented at the outset, is scarcely appreciated. As Plutarch says, "it is not possible to find in geometry more difficult and troublesome questions, or more simple and lucid explanations." But it is decidedly a rhetorical exaggeration when Plutarch goes on to say that we are deceived by the easiness of the successive steps into the belief that anyone could have discovered them for himself. On the contrary, the studied simplicity and the perfect finish of the treatises involve at the same time an element of mystery. Though each step depends upon the preceding ones, we are left in the dark as to how they were suggested to Archimedes. There is, in fact, much truth in a remark of Wallis to the effect that he seems "as it were of set purpose to have covered up the traces of his investigation as if he had grudged posterity the secret of his method of inquiry while he wished to extort from them assent to his results." Wallis adds with equal reason that not only Archimedes but nearly all the ancients so hid away from posterity their method of Analysis (though it is certain that they had one) that more modern mathematicians found it easier to invent a new Analysis than to seek out the old. This is no doubt the reason why Archimedes and other Greek geometers have received so little attention during the present century and why Archimedes is for the most part only vaguely remembered as the inventor of a screw, while even mathematicians scarcely know him except as the discoverer of the principle in hydrostatics which bears his name. It is only of recent years that we have had a satisfactory edition of the Greek text, that of Heiberg brought out in 1880–1, and I know of no complete translation since the German one of Nizze, published in 1824, which is now out of print and so rare that I had some difficulty in procuring a copy.

The plan of this work is then the same as that which I followed in editing the *Conics* of Apollonius. In this case, however, there has been less need as well as less opportunity for compression, and it has been possible to retain the numbering of the propositions and to enunciate them in a manner more nearly approaching the original without thereby making the enunciations obscure. Moreover, the subject matter is not so complicated as to necessitate absolute uniformity in the notation used (which is the only means whereby Apollonius can be made even tolerably readable), though I have tried to secure as much uniformity as was fairly possible. My main object has been to present a perfectly faithful reproduction of the treatises as they have come down to us, neither adding anything nor leaving out anything essential or important. The notes are for the most part intended to throw light on particular points in the text or to supply proofs of propositions assumed by Archimedes as known; sometimes I have thought it right to insert within square brackets after certain propositions, and in the same type, notes designed to bring out the exact significance of those propositions, in cases where to place such notes in the Introduction or at the bottom of the page might lead to their being overlooked.

Much of the Introduction is, as will be seen, historical; the rest is devoted partly to giving a more general view of certain methods employed by Archimedes and of their mathematical significance than would be possible in notes to separate propositions, and partly to the discussion of certain questions arising out of the subject matter upon which we have no positive historical data to guide us. In these latter cases, where it is necessary to put forward hypotheses for the purpose of explaining obscure points, I have been careful to call attention to their speculative character, though I have given the historical evidence where such can be quoted in support of a particular hypothesis, my object being to place side by side the authentic information which we possess and the inferences which have been or may be drawn from it, in order that the reader may be in a position to judge for himself how far he can accept the latter as probable. Perhaps I may be thought to owe an apology for the length of one chapter on the so-called νεύσεις, or *inclinationes*, which goes somewhat beyond what is necessary for the elucidation of Archimedes; but the subject is interesting, and I thought it well to make my account of it as complete as possible in order to round off, as it were, my studies in Apollonius and Archimedes.

PREFACE.

I have had one disappointment in preparing this book for the press. I was particularly anxious to place on or opposite the title-page a portrait of Archimedes, and I was encouraged in this idea by the fact that the title-page of Torelli's edition bears a representation in medallion form on which are endorsed the words *Archimedis effigies marmorea in veteri anaglypho Romae asservato*. Caution was however suggested when I found two more portraits wholly unlike this but still claiming to represent Archimedes, one of them appearing at the beginning of Peyrard's French translation of 1807, and the other in Gronovius' *Thesaurus Graecarum Antiquitatum*; and I thought it well to inquire further into the matter. I am now informed by Dr A. S. Murray of the British Museum that there does not appear to be any authority for any one of the three, and that writers on iconography apparently do not recognise an Archimedes among existing portraits. I was, therefore, reluctantly obliged to give up my idea.

The proof sheets have, as on the former occasion, been read over by my brother, Dr R. S. Heath, Principal of Mason College, Birmingham; and I desire to take this opportunity of thanking him for undertaking what might well have seemed, to any one less genuinely interested in Greek geometry, a thankless task.

T. L. HEATH.

March, 1897.

LIST OF THE PRINCIPAL WORKS CONSULTED.

JOSEPH TORELLI, *Archimedis quae supersunt omnia cum Eutocii Ascalonitae commentariis.* (Oxford, 1792.)

ERNST NIZZE, *Archimedes von Syrakus vorhandene Werke aus dem griechischen übersetzt und mit erläuternden und kritischen Anmerkungen begleitet.* (Stralsund, 1824.)

J. L. HEIBERG, *Archimedis opera omnia cum commentariis Eutocii.* (Leipzig, 1880–1.)

J. L. HEIBERG, *Quaestiones Archimedeae.* (Copenhagen, 1879.)

F. HULTSCH, Article *Archimedes* in Pauly-Wissowa's *Real-Encyclopädie der classischen Altertumswissenschaften.* (Edition of 1895, II. 1, pp. 507–539.)

C. A. BRETSCHNEIDER, *Die Geometrie und die Geometer vor Euklides.* (Leipzig, 1870.)

M. CANTOR, *Vorlesungen über Geschichte der Mathematik*, Band I, zweite Auflage. (Leipzig, 1894.)

G. FRIEDLEIN, *Procli Diadochi in primum Euclidis elementorum librum commentarii.* (Leipzig, 1873.)

JAMES GOW, *A short history of Greek Mathematics.* (Cambridge, 1884.)

SIEGMUND GÜNTHER, *Abriss der Geschichte der Mathematik und der Naturwissenschaften im Altertum* in Iwan von Müller's *Handbuch der klassischen Altertumswissenschaft,* v. 1.

HERMANN HANKEL, *Zur Geschichte der Mathematik in Alterthum und Mittelalter.* (Leipzig, 1874.)

J. L. HEIBERG, *Litterargeschichtliche Studien über Euklid.* (Leipzig, 1882.)

J. L. HEIBERG, *Euclidis elementa.* (Leipzig, 1883–8.)

F. HULTSCH, Article *Arithmetica* in Pauly-Wissowa's *Real-Encyclopädie,* II. 1, pp. 1066–1116.

LIST OF PRINCIPAL WORKS CONSULTED.

F. HULTSCH, *Heronis Alexandrini geometricorum et stereometricorum reliquiae.* (Berlin, 1864.)

F. HULTSCH, *Pappi Alexandrini collectionis quae supersunt.* (Berlin, 1876–8.)

GINO LORIA, *Il periodo aureo della geometria greca.* (Modena, 1895.)

MAXIMILIEN MARIE, *Histoire des sciences mathématiques et physiques,* Tome I. (Paris, 1883.)

J. H. T. MÜLLER, *Beiträge zur Terminologie der griechischen Mathematiker.* (Leipzig, 1860.)

G. H. F. NESSELMANN, *Die Algebra der Griechen.* (Berlin, 1842.)

F. SUSEMIHL, *Geschichte der griechischen Litteratur in der Alexandrinerzeit,* Band I. (Leipzig, 1891.)

P. TANNERY, *La Géométrie grecque,* Première partie, *Histoire générale de la Géométrie élémentaire.* (Paris, 1887.)

H. G. ZEUTHEN, *Die Lehre von den Kegelschnitten im Altertum.* (Copenhagen, 1886.)

H. G. ZEUTHEN, *Geschichte der Mathematik im Altertum und Mittelalter.* (Copenhagen, 1896.)

CONTENTS.

INTRODUCTION.

		PAGE
CHAPTER I.	ARCHIMEDES	xv
CHAPTER II.	MANUSCRIPTS AND PRINCIPAL EDITIONS—ORDER OF COMPOSITION—DIALECT—LOST WORKS .	xxiii
CHAPTER III.	RELATION OF ARCHIMEDES TO HIS PREDECESSORS	xxxix
§ 1.	Use of traditional geometrical methods . .	xl
§ 2.	Earlier discoveries affecting quadrature and cubature	xlvii
§ 3.	Conic Sections	lii
§ 4.	Surfaces of the second degree	liv
§ 5.	Two mean proportionals in continued proportion	lxvii
CHAPTER IV.	ARITHMETIC IN ARCHIMEDES	lxviii
§ 1.	Greek numeral system	lxix
§ 2.	Addition and subtraction	lxxi
§ 3.	Multiplication	lxxii
§ 4.	Division	lxxiii
§ 5.	Extraction of the square root	lxxiv
§ 6.	Early investigations of surds or incommensurables	lxxvii
§ 7.	Archimedes' approximations to $\sqrt{3}$ . . .	lxxx
§ 8.	Archimedes' approximations to the square roots of large numbers which are not complete squares	lxxxiv
	Note on alternative hypotheses with regard to the approximations to $\sqrt{3}$	xc

xiv CONTENTS.

		PAGE
CHAPTER V.	ON THE PROBLEMS KNOWN AS ΝΕΥΣΕΙΣ	c
§ 1.	Νεύσεις referred to by Archimedes	c
§ 2.	Mechanical constructions: the *conchoid* of Nicomedes	cv
§ 3.	Pappus' solution of the νεῦσις referred to in Props. 8, 9 *On Spirals*	cvii
§ 4.	The problem of the two mean proportionals	cx
§ 5.	The trisection of an angle	cxi
§ 6.	On certain *plane* νεύσεις	cxiii
CHAPTER VI.	CUBIC EQUATIONS	cxxiii
CHAPTER VII.	ANTICIPATIONS BY ARCHIMEDES OF THE INTEGRAL CALCULUS	cxlii
CHAPTER VIII.	THE TERMINOLOGY OF ARCHIMEDES	clv

THE WORKS OF ARCHIMEDES.

ON THE SPHERE AND CYLINDER, BOOK I.	1
,, ,, ,, ,, BOOK II.	56
MEASUREMENT OF A CIRCLE	91
ON CONOIDS AND SPHEROIDS	99
ON SPIRALS	151
ON THE EQUILIBRIUM OF PLANES, BOOK I.	189
,, ,, ,, ,, BOOK II.	203
THE SAND-RECKONER	221
QUADRATURE OF THE PARABOLA	233
ON FLOATING BODIES, BOOK I.	253
,, ,, ,, BOOK II.	263
BOOK OF LEMMAS	301
THE CATTLE-PROBLEM	319

ON THE SPHERE AND CYLINDER.

BOOK I.

"ARCHIMEDES to Dositheus greeting.

On a former occasion I sent you the investigations which I had up to that time completed, including the proofs, showing that any segment bounded by a straight line and a section of a right-angled cone [a parabola] is four-thirds of the triangle which has the same base with the segment and equal height. Since then certain theorems not hitherto demonstrated (ἀνελέγκτων) have occurred to me, and I have worked out the proofs of them. They are these: first, that the surface of any sphere is four times its greatest circle (τοῦ μεγίστου κύκλου); next, that the surface of any segment of a sphere is equal to a circle whose radius (ἡ ἐκ τοῦ κέντρου) is equal to the straight line drawn from the vertex (κορυφή) of the segment to the circumference of the circle which is the base of the segment; and, further, that any cylinder having its base equal to the greatest circle of those in the sphere, and height equal to the diameter of the sphere, is itself [*i.e.* in content] half as large again as the sphere, and its surface also [including its bases] is half as large again as the surface of the sphere. Now these properties were all along naturally inherent in the figures referred to (αὐτῇ τῇ φύσει προυπῆρχεν περὶ τὰ εἰρημένα σχήματα), but remained unknown to those who were before my time engaged in the study of geometry. Having, however, now discovered that the properties are true of these figures, I cannot feel any hesitation in setting them side by side both with my former investigations and with those of the theorems of Eudoxus on solids which are held to be most irrefragably established, namely, that any pyramid is one third part of the prism which has the same base with the pyramid and equal height, and that any cone is one third part of the cylinder which has the same base with the cone and equal height. For, though these properties also were naturally inherent in the figures all along, yet they were in fact unknown to all the many able geometers who lived before Eudoxus, and had not been observed by any one. Now, however, it will be open to those who possess the requisite ability to examine these discoveries of mine. They ought to have been published while Conon was still alive, for I should conceive that he would best have been able to grasp them and to pronounce upon them the appropriate verdict; but, as I judge it well to communicate them to those who are conversant with mathematics, I send them to you with the proofs written out, which it will be open to mathematicians to examine. Farewell.

I first set out the axioms* and the assumptions which I have used for the proofs of my propositions.

DEFINITIONS.

1. There are in a plane certain terminated bent lines (καμπύλαι γραμμαὶ πεπερασμέναι)†, which either lie wholly on the same side of the straight lines joining their extremities, or have no part of them on the other side.

2. I apply the term **concave in the same direction** to a line such that, if any two points on it are taken, either all the straight lines connecting the points fall on the same side of the line, or some fall on one and the same side while others fall on the line itself, but none on the other side.

---

* Though the word used is ἀξιώματα, the "axioms" are more of the nature of definitions; and in fact Eutocius in his notes speaks of them as such (ὅροι).

† Under the term *bent line* Archimedes includes not only curved lines of continuous curvature, but lines made up of any number of lines which may be either straight or curved.

ON THE SPHERE AND CYLINDER I. 3

3. Similarly also there are certain terminated surfaces, not themselves being in a plane but having their extremities in a plane, and such that they will either be wholly on the same side of the plane containing their extremities, or have no part of them on the other side.

4. I apply the term **concave in the same direction** to surfaces such that, if any two points on them are taken, the straight lines connecting the points either all fall on the same side of the surface, or some fall on one and the same side of it while some fall upon it, but none on the other side.

5. I use the term **solid sector**, when a cone cuts a sphere, and has its apex at the centre of the sphere, to denote the figure comprehended by the surface of the cone and the surface of the sphere included within the cone.

6. I apply the term **solid rhombus**, when two cones with the same base have their apices on opposite sides of the plane of the base in such a position that their axes lie in a straight line, to denote the solid figure made up of both the cones.

ASSUMPTIONS.

1. *Of all lines which have the same extremities the straight line is the least*[*].

[*] This well-known Archimedean assumption is scarcely, as it stands, a *definition* of a straight line, though Proclus says [p. 110 ed. Friedlein] "Archimedes defined (ὡρίσατο) the straight line as the least of those [lines] which have the same extremities. For because, as Euclid's definition says, ἐξ ἴσου κεῖται τοῖς ἐφ' ἑαυτῆς σημείοις, it is in consequence the least of those which have the same extremities." Proclus had just before [p. 109] explained Euclid's definition, which, as will be seen, is different from the ordinary version given in our text-books; a straight line is not "that which lies evenly between its extreme points," but "that which ἐξ ἴσου τοῖς ἐφ' ἑαυτῆς σημείοις κεῖται." The words of Proclus are, "He [Euclid] shows by means of this that the straight line alone [of all lines] occupies a distance (κατέχειν διάστημα) equal to that between the points on it. For, as far as one of its points is removed from another, so great is the length (μέγεθος) of the straight line of which the points are the extremities; and this is the meaning of τὸ ἐξ ἴσου κεῖσθαι τοῖς ἐφ' ἑαυτῆς σημείοις. But, if you take two points on a circumference or any other line, the distance cut off between them along the line is greater than the interval separating them; and this is the case with every line except the straight line." It appears then from this that Euclid's definition should be understood in a sense very like that of

2. Of other lines in a plane and having the same extremities, [any two] such are unequal whenever both are concave in the same direction and one of them is either wholly included between the other and the straight line which has the same extremities with it, or is partly included by, and is partly common with, the other; and that [line] which is included is the lesser [of the two].

3. Similarly, of surfaces which have the same extremities, if those extremities are in a plane, the plane is the least [in area].

4. Of other surfaces with the same extremities, the extremities being in a plane, [any two] such are unequal whenever both are concave in the same direction and one surface is either wholly included between the other and the plane which has the same extremities with it, or is partly included by, and partly common with, the other; and that [surface] which is included is the lesser [of the two in area].

5. Further, of unequal lines, unequal surfaces, and unequal solids, the greater exceeds the less by such a magnitude as, when added to itself, can be made to exceed any assigned magnitude among those which are comparable with [it and with] one another*.

These things being premised, *if a polygon be inscribed in a circle, it is plain that the perimeter of the inscribed polygon is less than the circumference of the circle;* for each of the sides of the polygon is less than that part of the circumference of the circle which is cut off by it."

Archimedes' assumption, and we might perhaps translate as follows, "A straight line is that which extends equally (ἐξ ἴσου κεῖται) with the points on it," or, to follow Proclus' interpretation more closely, "A straight line is that which represents equal extension with [the distances separating] the points on it."

* With regard to this assumption compare the Introduction, chapter III. § 2.

Proposition 1.

If a polygon be circumscribed about a circle, the perimeter of the circumscribed polygon is greater than the perimeter of the circle.

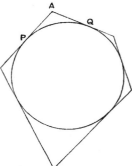

Let any two adjacent sides, meeting in A, touch the circle at P, Q respectively.

Then [*Assumptions*, 2]

$$PA + AQ > (\text{arc } PQ).$$

A similar inequality holds for each angle of the polygon; and, by addition, the required result follows.

Proposition 2.

Given two unequal magnitudes, it is possible to find two unequal straight lines such that the greater straight line has to the less a ratio less than the greater magnitude has to the less.

Let AB, D represent the two unequal magnitudes, AB being the greater.

Suppose BC measured along BA equal to D, and let GH be any straight line.

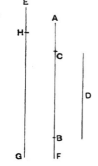

Then, if CA be added to itself a sufficient number of times, the sum will exceed D. Let AF be this sum, and take E on GH produced such that GH is the same multiple of HE that AF is of AC.

Thus $EH : HG = AC : AF$.

But, since $AF > D$ (or CB),

$$AC : AF < AC : CB.$$

Therefore, *componendo*,

$$EG : GH < AB : D.$$

Hence EG, GH are two lines satisfying the given condition.

Proposition 3.

Given two unequal magnitudes and a circle, it is possible to inscribe a polygon in the circle and to describe another about it so that the side of the circumscribed polygon may have to the side of the inscribed polygon a ratio less than that of the greater magnitude to the less.

Let A, B represent the given magnitudes, A being the greater.

Find [Prop. 2] two straight lines F, KL, of which F is the greater, such that
$$F : KL < A : B \quad\quad\quad\quad\quad\quad(1).$$

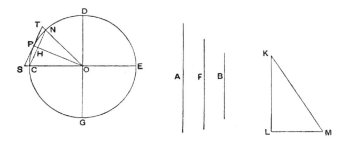

Draw LM perpendicular to LK and of such length that $KM = F$.

In the given circle let CE, DG be two diameters at right angles. Then, bisecting the angle DOC, bisecting the half again, and so on, we shall arrive ultimately at an angle (as NOC) less than twice the angle LKM.

Join NC, which (by the construction) will be the side of a regular polygon inscribed in the circle. Let OP be the radius of the circle bisecting the angle NOC (and therefore bisecting NC at right angles, in H, say), and let the tangent at P meet OC, ON produced in S, T respectively.

Now, since $\quad \angle CON < 2 \angle LKM$,

$\angle HOC < \angle LKM$,

and the angles at H, L are right;

therefore $MK : LK > OC : OH$
$> OP : OH$.

Hence $ST : CN < MK : LK$
$< F : LK$;

therefore, *a fortiori*, by (1),

$ST : CN < A : B$.

Thus two polygons are found satisfying the given condition.

Proposition 4.

Again, given two unequal magnitudes and a sector, it is possible to describe a polygon about the sector and to inscribe another in it so that the side of the circumscribed polygon may have to the side of the inscribed polygon a ratio less than the greater magnitude has to the less.

[The "inscribed polygon" found in this proposition is one which has for two sides the two radii bounding the sector, while the remaining sides (the number of which is, by construction, some power of 2) subtend equal parts of the arc of the sector; the "circumscribed polygon" is formed by the tangents parallel to the sides of the inscribed polygon and by the two bounding radii produced.]

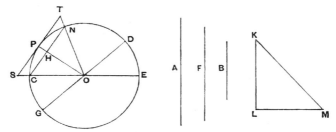

In this case we make the same construction as in the last proposition except that we bisect the angle COD of the sector, instead of the right angle between two diameters, then bisect the half again, and so on. The proof is exactly similar to the preceding one.

Proposition 5.

Given a circle and two unequal magnitudes, to describe a polygon about the circle and inscribe another in it, so that the circumscribed polygon may have to the inscribed a ratio less than the greater magnitude has to the less.

Let A be the given circle and B, C the given magnitudes, B being the greater.

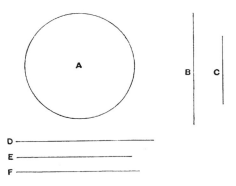

Take two unequal straight lines D, E, of which D is the greater, such that $D : E < B : C$ [Prop. 2], and let F be a mean proportional between D, E, so that D is also greater than F.

Describe (in the manner of Prop. 3) one polygon about the circle, and inscribe another in it, so that the side of the former has to the side of the latter a ratio less than the ratio $D : F$.

Thus the duplicate ratio of the side of the former polygon to the side of the latter is less than the ratio $D^2 : F^2$.

But the said duplicate ratio of the sides is equal to the ratio of the areas of the polygons, since they are similar;

therefore the area of the circumscribed polygon has to the area of the inscribed polygon a ratio less than the ratio $D^2 : F^2$, or $D : E$, and *a fortiori* less than the ratio $B : C$.

Proposition 6.

"Similarly we can show that, *given two unequal magnitudes and a sector, it is possible to circumscribe a polygon about the sector and inscribe in it another similar one so that the circumscribed may have to the inscribed a ratio less than the greater magnitude has to the less.*

And it is likewise clear that, *if a circle or a sector, as well as a certain area, be given, it is possible, by inscribing regular polygons in the circle or sector, and by continually inscribing such in the remaining segments, to leave segments of the circle or sector which are [together] less than the given area.* For this is proved in the *Elements* [Eucl. XII. 2].

But it is yet to be proved that, *given a circle or sector and an area, it is possible to describe a polygon about the circle or sector, such that the area remaining between the circumference and the circumscribed figure is less than the given area.*"

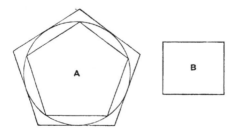

The proof for the circle (which, as Archimedes says, can be equally applied to a sector) is as follows.

Let A be the given circle and B the given area.

Now, there being two unequal magnitudes $A + B$ and A, let a polygon (C) be circumscribed about the circle and a polygon (I) inscribed in it [as in Prop. 5], so that

$$C : I < A + B : A \quad \ldots\ldots\ldots\ldots\ldots\ldots (1).$$

The circumscribed polygon (C) shall be that required.

For the circle (A) is greater than the inscribed polygon (I).

Therefore, from (1), *a fortiori*,

$$C : A < A + B : A,$$

whence $\qquad C < A + B,$

or $\qquad C - A < B.$

Proposition 7.

If in an isosceles cone [i.e. *a right circular cone*] *a pyramid be inscribed having an equilateral base, the surface of the pyramid excluding the base is equal to a triangle having its base equal to the perimeter of the base of the pyramid and its height equal to the perpendicular drawn from the apex on one side of the base.*

Since the sides of the base of the pyramid are equal, it follows that the perpendiculars from the apex to all the sides of the base are equal; and the proof of the proposition is obvious.

Proposition 8.

If a pyramid be circumscribed about an isosceles cone, the surface of the pyramid excluding its base is equal to a triangle having its base equal to the perimeter of the base of the pyramid and its height equal to the side [i.e. *a generator*] *of the cone.*

The base of the pyramid is a polygon circumscribed about the circular base of the cone, and the line joining the apex of the cone or pyramid to the point of contact of any side of the polygon is perpendicular to that side. Also all these perpendiculars, being generators of the cone, are equal; whence the proposition follows immediately.

Proposition 9.

If in the circular base of an isosceles cone a chord be placed, and from its extremities straight lines be drawn to the apex of the cone, the triangle so formed will be less than the portion of the surface of the cone intercepted between the lines drawn to the apex.

Let ABC be the circular base of the cone, and O its apex.

Draw a chord AB in the circle, and join OA, OB. Bisect the arc ACB in C, and join AC, BC, OC.

Then $\triangle OAC + \triangle OBC > \triangle OAB.$

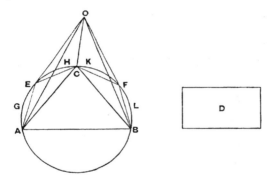

Let the excess of the sum of the first two triangles over the third be equal to the area D.

Then D is either less than the sum of the segments AEC, CFB, or not less.

I. Let D be not less than the sum of the segments referred to.

We have now two surfaces

(1) that consisting of the portion $OAEC$ of the surface of the cone together with the segment AEC, and

(2) the triangle OAC;

and, since the two surfaces have the same extremities (the perimeter of the triangle OAC), the former surface is greater than the latter, which is *included* by it [*Assumptions*, 3 or 4].

Hence (surface $OAEC$) + (segment AEC) > $\triangle\, OAC$.

Similarly (surface $OCFB$) + (segment CFB) > $\triangle\, OBC$.

Therefore, since D is not less than the sum of the segments, we have, by addition,

(surface $OAECFB$) + D > $\triangle\, OAC + \triangle\, OBC$

$> \triangle\, OAB + D$, by hypothesis.

Taking away the common part D, we have the required result.

II. Let D be less than the sum of the segments AEC, CFB.

If now we bisect the arcs AC, CB, then bisect the halves, and so on, we shall ultimately leave segments which are together less than D. [Prop. 6]

Let AGE, EHC, CKF, FLB be those segments, and join OE, OF.

Then, as before,

(surface $OAGE$) + (segment AGE) > $\triangle\, OAE$

and (surface $OEHC$) + (segment EHC) > $\triangle\, OEC$.

Therefore (surface $OAGHC$) + (segments AGE, EHC)

$> \triangle\, OAE + \triangle\, OEC$

$> \triangle\, OAC$, *a fortiori*.

Similarly for the part of the surface of the cone bounded by OC, OB and the arc CFB.

Hence, by addition,

(surface $OAGEHCKFLB$) + (segments AGE, EHC, CKF, FLB)

$> \triangle\, OAC + \triangle\, OBC$

$> \triangle\, OAB + D$, by hypothesis.

But the sum of the segments is less than D, and the required result follows.

Proposition 10.

If in the plane of the circular base of an isosceles cone two tangents be drawn to the circle meeting in a point, and the points of contact and the point of concourse of the tangents be respectively joined to the apex of the cone, the sum of the two triangles formed by the joining lines and the two tangents are together greater than the included portion of the surface of the cone.

Let ABC be the circular base of the cone, O its apex, AD, BD the two tangents to the circle meeting in D. Join OA, OB, OD.

Let ECF be drawn touching the circle at C, the middle point of the arc ACB, and therefore parallel to AB. Join OE, OF.

Then $$ED + DF > EF,$$
and, adding $AE + FB$ to each side,
$$AD + DB > AE + EF + FB.$$

Now OA, OC, OB, being generators of the cone, are equal, and they are respectively perpendicular to the tangents at A, C, B.

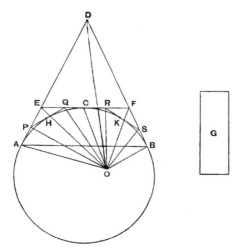

It follows that

$$\triangle OAD + \triangle ODB > \triangle OAE + \triangle OEF + \triangle OFB.$$

Let the area G be equal to the excess of the first sum over the second.

G is then either less, or not less, than the sum of the spaces $EAHC$, $FCKB$ remaining between the circle and the tangents, which sum we will call L.

I. Let G be not less than L.

We have now two surfaces

(1) that of the pyramid with apex O and base $AEFB$, excluding the face OAB,

(2) that consisting of the part $OACB$ of the surface of the cone together with the segment ACB.

These two surfaces have the same extremities, viz. the perimeter of the triangle OAB, and, since the former *includes* the latter, the former is the greater [*Assumptions*, 4].

That is, the surface of the pyramid exclusive of the face OAB is greater than the sum of the surface $OACB$ and the segment ACB.

Taking away the segment from each sum, we have

$$\triangle OAE + \triangle OEF + \triangle OFB + L > \text{the surface } OAHCKB.$$

And G is not less than L.

It follows that

$$\triangle OAE + \triangle OEF + \triangle OFB + G,$$

which is by hypothesis equal to $\triangle OAD + \triangle ODB$, is greater than the same surface.

II. Let G be less than L.

If we bisect the arcs AC, CB and draw tangents at their middle points, then bisect the halves and draw tangents, and so on, we shall lastly arrive at a polygon such that the sum of the parts remaining between the sides of the polygon and the circumference of the segment is less than G.

Let the remainders be those between the segment and the polygon $APQRSB$, and let their sum be M. Join OP, OQ, etc.

Then, as before,
$$\triangle OAE + \triangle OEF + \triangle OFB > \triangle OAP + \triangle OPQ + \ldots + \triangle OSB.$$
Also, as before,
(surface of pyramid $OAPQRSB$ excluding the face OAB)
> the part $OACB$ of the surface of the cone together with the segment ACB.

Taking away the segment from each sum,
$$\triangle OAP + \triangle OPQ + \ldots + M > \text{the part } OACB \text{ of the}$$
surface of the cone.

Hence, *a fortiori*,
$$\triangle OAE + \triangle OEF + \triangle OFB + G,$$
which is by hypothesis equal to
$$\triangle OAD + \triangle ODB,$$
is greater than the part $OACB$ of the surface of the cone.

Proposition 11.

If a plane parallel to the axis of a right cylinder cut the cylinder, the part of the surface of the cylinder cut off by the plane is greater than the area of the parallelogram in which the plane cuts it.

Proposition 12.

If at the extremities of two generators of any right cylinder tangents be drawn to the circular bases in the planes of those bases respectively, and if the pairs of tangents meet, the parallelograms formed by each generator and the two corresponding tangents respectively are together greater than the included portion of the surface of the cylinder between the two generators.

[The proofs of these two propositions follow exactly the methods of Props. 9, 10 respectively, and it is therefore unnecessary to reproduce them.]

"From the properties thus proved it is clear (1) that, *if a pyramid be inscribed in an isosceles cone, the surface of the pyramid excluding the base is less than the surface of the cone* [*excluding the base*], and (2) that, *if a pyramid be circumscribed about an isosceles cone, the surface of the pyramid excluding the base is greater than the surface of the cone excluding the base.*

"It is also clear from what has been proved both (1) that, *if a prism be inscribed in a right cylinder, the surface of the prism made up of its parallelograms* [i.e. *excluding its bases*] *is less than the surface of the cylinder excluding its bases*, and (2) that, *if a prism be circumscribed about a right cylinder, the surface of the prism made up of its parallelograms is greater than the surface of the cylinder excluding its bases.*"

Proposition 13.

The surface of any right cylinder excluding the bases is equal to a circle whose radius is a mean proportional between the side [i.e. *a generator*] *of the cylinder and the diameter of its base.*

Let the base of the cylinder be the circle A, and make CD equal to the diameter of this circle, and EF equal to the height of the cylinder.

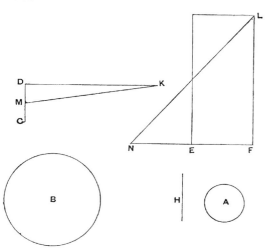

ON THE SPHERE AND CYLINDER I.

Let H be a mean proportional between CD, EF, and B a circle with radius equal to H.

Then the circle B shall be equal to the surface of the cylinder (excluding the bases), which we will call S.

For, if not, B must be either greater or less than S.

I. Suppose $B < S$.

Then it is possible to circumscribe a regular polygon about B, and to inscribe another in it, such that the ratio of the former to the latter is less than the ratio $S : B$.

Suppose this done, and circumscribe about A a polygon similar to that described about B; then erect on the polygon about A a prism of the same height as the cylinder. The prism will therefore be circumscribed to the cylinder.

Let KD, perpendicular to CD, and FL, perpendicular to EF, be each equal to the perimeter of the polygon about A. Bisect CD in M, and join MK.

Then $\triangle KDM =$ the polygon about A.

Also $\square EL =$ surface of prism (excluding bases).

Produce FE to N so that $FE = EN$, and join NL.

Now the polygons about A, B, being similar, are in the duplicate ratio of the radii of A, B.

Thus

$\triangle KDM :$ (polygon about B) $= MD^2 : H^2$
$\qquad = MD^2 : CD . EF$
$\qquad = MD : NF$
$\qquad = \triangle KDM : \triangle LFN$

(since $DK = FL$).

Therefore (polygon about B) $= \triangle LFN$
$\qquad = \square EL$
$\qquad =$ (surface of prism about A),

from above.

But (polygon about B) : (polygon in B) $< S : B$.

Therefore

(surface of prism about A) : (polygon in B) < S : B,

and, alternately,

(surface of prism about A) : S < (polygon in B) : B;

which is impossible, since the surface of the prism is greater than S, while the polygon inscribed in B is less than B.

Therefore $B \not< S$.

II. Suppose $B > S$.

Let a regular polygon be circumscribed about B and another inscribed in it so that

(polygon about B) : (polygon in B) < B : S.

Inscribe in A a polygon similar to that inscribed in B, and erect a prism on the polygon inscribed in A of the same height as the cylinder.

Again, let DK, FL, drawn as before, be each equal to the perimeter of the polygon inscribed in A.

Then, in this case,

$$\triangle KDM > \text{(polygon inscribed in } A\text{)}$$

(since the perpendicular from the centre on a side of the polygon is less than the radius of A).

Also $\triangle LFN = \square\, EL =$ surface of prism (excluding bases).

Now

(polygon in A) : (polygon in B) $= MD^2 : H^2$,

$\qquad\qquad\qquad = \triangle KDM : \triangle LFN$, as before.

And $\qquad\qquad \triangle KDM > $ (polygon in A).

Therefore

$\triangle LFN$, or (surface of prism) > (polygon in B).

But this is impossible, because

(polygon about B) : (polygon in B) < B : S,

$\qquad\qquad\qquad <$ (polygon about B) : S, *a fortiori*,

so that $\qquad$ (polygon in B) > S,

$\qquad\qquad\qquad >$ (surface of prism), *a fortiori*.

Hence B is neither greater nor less than S, and therefore
$$B = S.$$

Proposition 14.

The surface of any isosceles cone excluding the base is equal to a circle whose radius is a mean proportional between the side of the cone [*a generator*] *and the radius of the circle which is the base of the cone.*

Let the circle A be the base of the cone; draw C equal to the radius of the circle, and D equal to the side of the cone, and let E be a mean proportional between C, D.

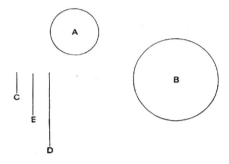

Draw a circle B with radius equal to E.

Then shall B be equal to the surface of the cone (excluding the base), which we will call S.

If not, B must be either greater or less than S.

I. Suppose $B < S$.

Let a regular polygon be described about B and a similar one inscribed in it such that the former has to the latter a ratio less than the ratio $S : B$.

Describe about A another similar polygon, and on it set up a pyramid with apex the same as that of the cone.

Then (polygon about A) : (polygon about B)
$= C^2 : E^2$
$= C : D$
$=$ (polygon about A) : (surface of pyramid excluding base).

Therefore

(surface of pyramid) = (polygon about B).

Now (polygon about B) : (polygon in B) < $S : B$.

Therefore

(surface of pyramid) : (polygon in B) < $S : B$,

which is impossible, (because the surface of the pyramid is greater than S, while the polygon in B is less than B).

Hence $B \not< S$.

II. Suppose $B > S$.

Take regular polygons circumscribed and inscribed to B such that the ratio of the former to the latter is less than the ratio $B : S$.

Inscribe in A a similar polygon to that inscribed in B, and erect a pyramid on the polygon inscribed in A with apex the same as that of the cone.

In this case

(polygon in A) : (polygon in B) = $C^2 : E^2$

$= C : D$

> (polygon in A) : (surface of pyramid excluding base).

This is clear because the ratio of C to D is greater than the ratio of the perpendicular from the centre of A on a side of the polygon to the perpendicular from the apex of the cone on the same side*.

Therefore

(surface of pyramid) > (polygon in B).

But (polygon about B) : (polygon in B) < $B : S$.

Therefore, *a fortiori*,

(polygon about B) : (surface of pyramid) < $B : S$;

which is impossible.

Since therefore B is neither greater nor less than S,

$B = S$.

---

* This is of course the geometrical equivalent of saying that, if a, β be two angles each less than a right angle, and $a > \beta$, then $\sin a > \sin \beta$.

Proposition 15.

The surface of any isosceles cone has the same ratio to its base as the side of the cone has to the radius of the base.

By Prop. 14, the surface of the cone is equal to a circle whose radius is a mean proportional between the side of the cone and the radius of the base.

Hence, since circles are to one another as the squares of their radii, the proposition follows.

Proposition 16.

If an isosceles cone be cut by a plane parallel to the base, the portion of the surface of the cone between the parallel planes is equal to a circle whose radius is a mean proportional between (1) *the portion of the side of the cone intercepted by the parallel planes and* (2) *the line which is equal to the sum of the radii of the circles in the parallel planes.*

Let OAB be a triangle through the axis of a cone, DE its intersection with the plane cutting off the frustum, and OFC the axis of the cone.

Then the surface of the cone OAB is equal to a circle whose radius is equal to $\sqrt{OA \cdot AC}$. [Prop. 14.]

Similarly the surface of the cone ODE is equal to a circle whose radius is equal to $\sqrt{OD \cdot DF}$.

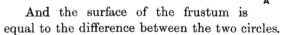

And the surface of the frustum is equal to the difference between the two circles.

Now

$$OA \cdot AC - OD \cdot DF = DA \cdot AC + OD \cdot AC - OD \cdot DF.$$

But $OD \cdot AC = OA \cdot DF$,

since $OA : AC = OD : DF.$

Hence $OA \cdot AC - OD \cdot DF = DA \cdot AC + DA \cdot DF$
$= DA \cdot (AC + DF)$.

And, since circles are to one another as the squares of their radii, it follows that the difference between the circles whose radii are $\sqrt{OA \cdot AC}$, $\sqrt{OD \cdot DF}$ respectively is equal to a circle whose radius is $\sqrt{DA \cdot (AC + DF)}$.

Therefore the surface of the frustum is equal to this circle.

Lemmas.

"1. *Cones having equal height have the same ratio as their bases; and those having equal bases have the same ratio as their heights*[*].

2. *If a cylinder be cut by a plane parallel to the base, then, as the cylinder is to the cylinder, so is the axis to the axis* [†].

3. *The cones which have the same bases as the cylinders* [*and equal height*] *are in the same ratio as the cylinders.*

4. *Also the bases of equal cones are reciprocally proportional to their heights; and those cones whose bases are reciprocally proportional to their heights are equal* [‡].

5. *Also the cones, the diameters of whose bases have the same ratio as their axes, are to one another in the triplicate ratio of the diameters of the bases* [§].

And all these propositions have been proved by earlier geometers."

---

[*] Euclid XII. 11. "Cones and cylinders of equal height are to one another as their bases."
Euclid XII. 14. "Cones and cylinders on equal bases are to one another as their heights."

[†] Euclid XII. 13. "If a cylinder be cut by a plane parallel to the opposite planes [the bases], then, as the cylinder is to the cylinder, so will the axis be to the axis."

[‡] Euclid XII. 15. "The bases of equal cones and cylinders are reciprocally proportional to their heights; and those cones and cylinders whose bases are reciprocally proportional to their heights are equal."

[§] Euclid XII. 12. "Similar cones and cylinders are to one another in the triplicate ratio of the diameters of their bases."

Proposition 17.

If there be two isosceles cones, and the surface of one cone be equal to the base of the other, while the perpendicular from the centre of the base [of the first cone] on the side of that cone is equal to the height [of the second], the cones will be equal.

Let OAB, DEF be triangles through the axes of two cones respectively, C, G the centres of the respective bases, GH the

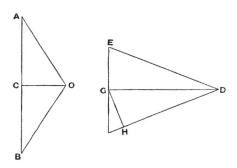

perpendicular from G on FD; and suppose that the base of the cone OAB is equal to the surface of the cone DEF, and that $OC = GH$.

Then, since the base of OAB is equal to the surface of DEF,

(base of cone OAB) : (base of cone DEF)

$\quad =$ (surface of DEF) : (base of DEF)

$\quad = DF : FG$ [Prop. 15]

$\quad = DG : GH$, by similar triangles,

$\quad = DG : OC$.

Therefore the bases of the cones are reciprocally proportional to their heights; whence the cones are equal. [*Lemma* 4.]

Proposition 18.

Any solid rhombus consisting of isosceles cones is equal to the cone which has its base equal to the surface of one of the cones composing the rhombus and its height equal to the perpendicular drawn from the apex of the second cone to one side of the first cone.

Let the rhombus be $OABD$ consisting of two cones with apices O, D and with a common base (the circle about AB as diameter).

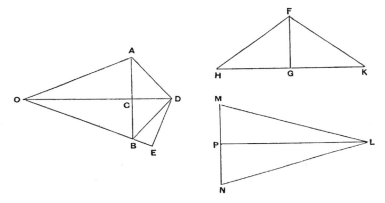

Let FHK be another cone with base equal to the surface of the cone OAB and height FG equal to DE, the perpendicular from D on OB.

Then shall the cone FHK be equal to the rhombus.

Construct a third cone LMN with base (the circle about MN) equal to the base of OAB and height LP equal to OD.

Then, since $\qquad LP = OD$,
$$LP : CD = OD : CD.$$
But [*Lemma* 1] $OD : CD =$ (rhombus $OADB$) : (cone DAB),
and $\qquad LP : CD =$ (cone LMN) : (cone DAB).

It follows that

(rhombus $OADB$) = (cone LMN) .... (1).

Again, since $AB = MN$, and

(surface of OAB) = (base of FHK),

(base of FHK) : (base of LMN)
$$= \text{(surface of } OAB\text{)} : \text{(base of } OAB\text{)}$$
$$= OB : BC \quad \text{[Prop. 15]}$$
$$= OD : DE, \text{ by similar triangles,}$$
$$= LP : FG, \text{ by hypothesis.}$$

Thus, in the cones FHK, LMN, the bases are reciprocally proportional to the heights.

Therefore the cones FHK, LMN are equal,

and hence, by (1), the cone FHK is equal to the given solid rhombus.

Proposition 19.

If an isosceles cone be cut by a plane parallel to the base, and on the resulting circular section a cone be described having as its apex the centre of the base [of the first cone], and if the rhombus so formed be taken away from the whole cone, the part remaining will be equal to the cone with base equal to the surface of the portion of the first cone between the parallel planes and with height equal to the perpendicular drawn from the centre of the base of the first cone on one side of that cone.

Let the cone OAB be cut by a plane parallel to the base in the circle on DE as diameter. Let C be the centre of the base of the cone, and with C as apex and the circle about DE as base describe a cone, making with the cone ODE the rhombus $ODCE$.

Take a cone FGH with base equal to the surface of the frustum $DABE$ and height equal to the perpendicular (CK) from C on AO.

Then shall the cone FGH be equal to the difference between the cone OAB and the rhombus $ODCE$.

Take (1) a cone LMN with base equal to the surface of the cone OAB, and height equal to CK,

(2) a cone PQR with base equal to the surface of the cone ODE and height equal to CK.

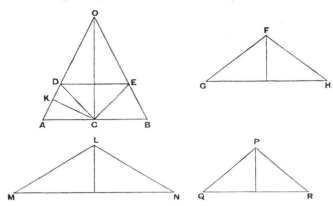

Now, since the surface of the cone OAB is equal to the surface of the cone ODE together with that of the frustum $DABE$, we have, by the construction,

(base of LMN) = (base of FGH) + (base of PQR)

and, since the heights of the three cones are equal,

(cone LMN) = (cone FGH) + (cone PQR).

But the cone LMN is equal to the cone OAB [Prop. 17], and the cone PQR is equal to the rhombus $ODCE$ [Prop. 18].

Therefore (cone OAB) = (cone FGH) + (rhombus $ODCE$), and the proposition is proved.

Proposition 20.

If one of the two isosceles cones forming a rhombus be cut by a plane parallel to the base and on the resulting circular section a cone be described having the same apex as the second cone, and if the resulting rhombus be taken from the whole rhombus, the remainder will be equal to the cone with base equal to the surface of the portion of the cone between the parallel planes and with height equal to the perpendicular drawn from the apex of the second* cone to the side of the first cone.

---

* There is a slight error in Heiberg's translation "prioris coni" and in the corresponding note, p. 93. The perpendicular is not drawn from the apex of the cone which is cut by the plane but from the apex of the other.

ON THE SPHERE AND CYLINDER I.

Let the rhombus be $OACB$, and let the cone OAB be cut by a plane parallel to its base in the circle about DE as diameter. With this circle as base and C as apex describe a cone, which therefore with ODE forms the rhombus $ODCE$.

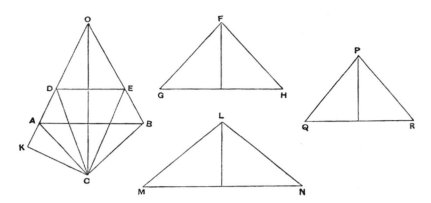

Take a cone FGH with base equal to the surface of the frustum $DABE$ and height equal to the perpendicular (CK) from C on OA.

The cone FGH shall be equal to the difference between the rhombi $OACB$, $ODCE$.

For take (1) a cone LMN with base equal to the surface of OAB and height equal to CK,

(2) a cone PQR, with base equal to the surface of ODE, and height equal to CK.

Then, since the surface of OAB is equal to the surface of ODE together with that of the frustum $DABE$, we have, by construction,

(base of LMN) = (base of PQR) + (base of FGH),

and the three cones are of equal height;

therefore (cone LMN) = (cone PQR) + (cone FGH).

But the cone LMN is equal to the rhombus $OACB$, and the cone PQR is equal to the rhombus $ODCE$ [Prop. 18].

Hence the cone FGH is equal to the difference between the two rhombi $OACB$, $ODCE$.

Proposition 21.

A regular polygon of an even number of sides being inscribed in a circle, as $ABC...A'...C'B'A$, so that AA' is a diameter, if two angular points next but one to each other, as B, B', be joined, and the other lines parallel to BB' and joining pairs of angular points be drawn, as CC', $DD'...$, then

$$(BB' + CC' + ...) : AA' = A'B : BA.$$

Let BB', CC', DD',... meet AA' in F, G, H,...; and let CB', DC',... be joined meeting AA' in K, L,... respectively.

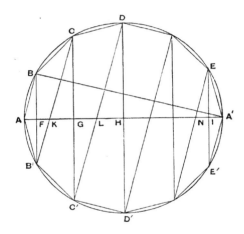

Then clearly CB', DC',... are parallel to one another and to AB.

Hence, by similar triangles,

$$BF : FA = B'F : FK$$
$$= CG : GK$$
$$= C'G : GL$$
$$\dots\dots\dots\dots$$
$$= E'I : IA';$$

and, summing the antecedents and consequents respectively, we have
$$(BB' + CC' + \ldots) : AA' = BF : FA$$
$$= A'B : BA.$$

Proposition 22.

If a polygon be inscribed in a segment of a circle LAL' so that all its sides excluding the base are equal and their number even, as $LK\ldots A\ldots K'L'$, A being the middle point of the segment, and if the lines BB', CC',... parallel to the base LL' and joining pairs of angular points be drawn, then
$$(BB' + CC' + \ldots + LM) : AM = A'B : BA,$$
where M is the middle point of LL' and AA' is the diameter through M.

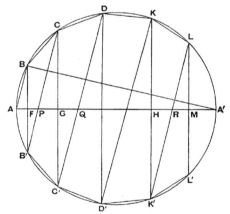

Joining CB', DC',...LK', as in the last proposition, and supposing that they meet AM in P, Q,...R, while BB', CC',..., KK' meet AM in F, G,... H, we have, by similar triangles,
$$BF : FA = B'F : FP$$
$$= CG : PG$$
$$= C'G : GQ$$
$$\ldots\ldots\ldots\ldots$$
$$= LM : RM;$$

and, summing the antecedents and consequents, we obtain
$$(BB' + CC' + \ldots + LM) : AM = BF : FA$$
$$= A'B : BA.$$

Proposition 23.

Take a great circle $ABC\ldots$ of a sphere, and inscribe in it a regular polygon whose sides are a multiple of four in number. Let AA', MM' be diameters at right angles and joining opposite angular points of the polygon.

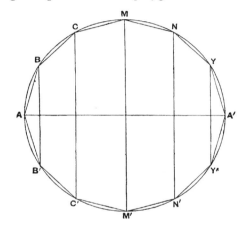

Then, if the polygon and great circle revolve together about the diameter AA', the angular points of the polygon, except A, A', will describe circles on the surface of the sphere at right angles to the diameter AA'. Also the sides of the polygon will describe portions of conical surfaces, e.g. BC will describe a surface forming part of a cone whose base is a circle about CC' as diameter and whose apex is the point in which CB, $C'B'$ produced meet each other and the diameter AA'.

Comparing the hemisphere MAM' and that half of the figure described by the revolution of the polygon which is included in the hemisphere, we see that the surface of the hemisphere and the surface of the inscribed figure have the same boundaries in one plane (viz. the circle on MM' as diameter), the former surface entirely includes the latter, and they are both concave in the same direction.

Therefore [*Assumptions*, 4] the surface of the hemisphere is greater than that of the inscribed figure; and the same is true of the other halves of the figures.

Hence *the surface of the sphere is greater than the surface described by the revolution of the polygon inscribed in the great circle about the diameter of the great circle.*

Proposition 24.

If a regular polygon $AB...A'...B'A$, the number of whose sides is a multiple of four, be inscribed in a great circle of a sphere, and if BB' subtending two sides be joined, and all the other lines parallel to BB' and joining pairs of angular points be drawn, then the surface of the figure inscribed in the sphere by the revolution of the polygon about the diameter AA' is equal to a circle the square of whose radius is equal to the rectangle

$$BA\,(BB' + CC' + ...).$$

The surface of the figure is made up of the surfaces of parts of different cones.

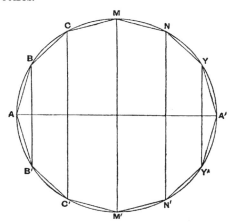

Now the surface of the cone ABB' is equal to a circle whose radius is $\sqrt{BA \cdot \tfrac{1}{2}BB'}$. [Prop. 14]

The surface of the frustum $BB'C'C$ is equal to a circle of radius $\sqrt{BC \cdot \tfrac{1}{2}(BB' + CC')}$, [Prop. 16] and so on.

It follows, since $BA = BC = ...$, that the whole surface is equal to a circle whose radius is equal to
$$\sqrt{BA\,(BB' + CC' + ... + MM' + ... + YY')}.$$

Proposition 25.

The surface of the figure inscribed in a sphere as in the last propositions, consisting of portions of conical surfaces, is less than four times the greatest circle in the sphere.

Let $AB...A'...B'A$ be a regular polygon inscribed in a great circle, the number of its sides being a multiple of four.

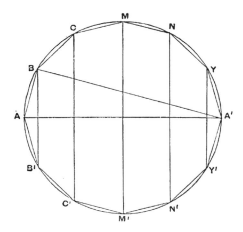

As before, let BB' be drawn subtending two sides, and $CC',...YY'$ parallel to BB'.

Let R be a circle such that the square of its radius is equal to
$$AB\,(BB' + CC' + ... + YY'),$$
so that the surface of the figure inscribed in the sphere is equal to R. [Prop. 24]

Now
$$(BB' + CC' + \ldots + YY') : AA' = A'B : AB, \quad [\text{Prop. 21}]$$
whence $AB(BB' + CC' + \ldots + YY') = AA' \cdot A'B$.

Hence (radius of R)$^2 = AA' \cdot A'B$

$< AA'^2$.

Therefore the surface of the inscribed figure, or the circle R, is less than four times the circle $AMA'M'$.

Proposition 26.

The figure inscribed as above in a sphere is equal [in volume] to a cone whose base is a circle equal to the surface of the figure inscribed in the sphere and whose height is equal to the perpendicular drawn from the centre of the sphere to one side of the polygon.

Suppose, as before, that $AB \ldots A' \ldots B'A$ is the regular polygon inscribed in a great circle, and let BB', CC', ... be joined.

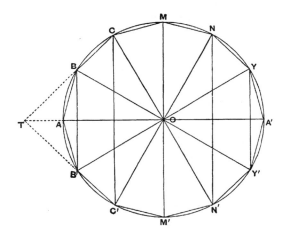

With apex O construct cones whose bases are the circles on BB', CC', ... as diameters in planes perpendicular to AA'.

Then $OBAB'$ is a solid rhombus, and its volume is equal to a cone whose base is equal to the surface of the cone ABB' and whose height is equal to the perpendicular from O on AB [Prop. 18]. Let the length of the perpendicular be p.

Again, if CB, $C'B'$ produced meet in T, the portion of the solid figure which is described by the revolution of the triangle BOC about AA' is equal to the difference between the rhombi $OCTC'$ and $OBTB'$, i.e. to a cone whose base is equal to the surface of the frustum $BB'C'C$ and whose height is p [Prop. 20].

Proceeding in this manner, and adding, we prove that, since cones of equal height are to one another as their bases, the volume of the solid of revolution is equal to a cone with height p and base equal to the sum of the surfaces of the cone BAB', the frustum $BB'C'C$, etc., i.e. a cone with height p and base equal to the surface of the solid.

Proposition 27.

The figure inscribed in the sphere as before is less than four times the cone whose base is equal to a great circle of the sphere and whose height is equal to the radius of the sphere.

By Prop. 26 the volume of the solid figure is equal to a cone whose base is equal to the surface of the solid and whose height is p, the perpendicular from O on any side of the polygon. Let R be such a cone.

Take also a cone S with base equal to the great circle, and height equal to the radius, of the sphere.

Now, since the surface of the inscribed solid is less than four times the great circle [Prop. 25], the base of the cone R is less than four times the base of the cone S.

Also the height (p) of R is less than the height of S.

Therefore the volume of R is less than four times that of S; and the proposition is proved.

Proposition 28.

Let a regular polygon, whose sides are a multiple of four in number, be circumscribed about a great circle of a given sphere, as $AB...A'...B'A$; and about the polygon describe another circle, which will therefore have the same centre as the great circle of the sphere. Let AA' bisect the polygon and cut the sphere in a, a'.

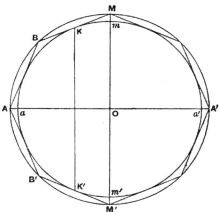

If the great circle and the circumscribed polygon revolve together about AA', the great circle will describe the surface of a sphere, the angular points of the polygon except A, A' will move round the surface of a larger sphere, the points of contact of the sides of the polygon with the great circle of the inner sphere will describe circles on that sphere in planes perpendicular to AA', and the sides of the polygon themselves will describe portions of conical surfaces. *The circumscribed figure will thus be greater than the sphere itself.*

Let any side, as BM, touch the inner circle in K, and let K' be the point of contact of the circle with $B'M'$.

Then the circle described by the revolution of KK' about AA' is the boundary in one plane of two surfaces

(1) the surface formed by the revolution of the circular segment KaK', and

(2) the surface formed by the revolution of the part $KB...A...B'K'$ of the polygon.

Now the second surface entirely includes the first, and they are both concave in the same direction;

therefore [*Assumptions*, 4] the second surface is greater than the first.

The same is true of the portion of the surface on the opposite side of the circle on KK' as diameter.

Hence, adding, we see that *the surface of the figure circumscribed to the given sphere is greater than that of the sphere itself.*

Proposition 29.

In a figure circumscribed to a sphere in the manner shown in the previous proposition the surface is equal to a circle the square on whose radius is equal to $AB(BB' + CC' + ...)$.

For the figure circumscribed to the sphere is inscribed in a larger sphere, and the proof of Prop. 24 applies.

Proposition 30.

The surface of a figure circumscribed as before about a sphere is greater than four times the great circle of the sphere.

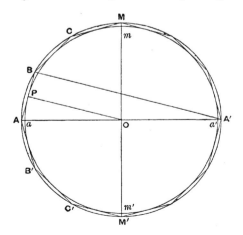

Let $AB...A'...B'A$ be the regular polygon of $4n$ sides which by its revolution about AA' describes the figure circumscribing the sphere of which $ama'm'$ is a great circle. Suppose aa', AA' to be in one straight line.

Let R be a circle equal to the surface of the circumscribed solid.

Now $(BB' + CC' + ...) : AA' = A'B : BA$, [as in Prop. 21]
so that $AB(BB' + CC' + ...) = AA' \cdot A'B$.
Hence (radius of R) $= \sqrt{AA' \cdot A'B}$ [Prop. 29]
$> A'B$.

But $A'B = 2OP$, where P is the point in which AB touches the circle $ama'm'$.

Therefore (radius of R) $>$ (diameter of circle $ama'm'$);

whence R, and therefore the surface of the circumscribed solid, is greater than four times the great circle of the given sphere.

Proposition 31.

The solid of revolution circumscribed as before about a sphere is equal to a cone whose base is equal to the surface of the solid and whose height is equal to the radius of the sphere.

The solid is, as before, a solid inscribed in a larger sphere; and, since the perpendicular on any side of the revolving polygon is equal to the radius of the inner sphere, the proposition is identical with Prop. 26.

COR. *The solid circumscribed about the smaller sphere is greater than four times the cone whose base is a great circle of the sphere and whose height is equal to the radius of the sphere.*

For, since the surface of the solid is greater than four times the great circle of the inner sphere [Prop. 30], the cone whose base is equal to the surface of the solid and whose height is the radius of the sphere is greater than four times the cone of the same height which has the great circle for base. [*Lemma* 1.]

Hence, by the proposition, the volume of the solid is greater than four times the latter cone.

Proposition 32.

If a regular polygon with $4n$ sides be inscribed in a great circle of a sphere, as $ab...a'...b'a$, and a similar polygon $AB...A'...B'A$ be described about the great circle, and if the polygons revolve with the great circle about the diameters aa', AA' respectively, so that they describe the surfaces of solid figures inscribed in and circumscribed to the sphere respectively, then

(1) *the surfaces of the circumscribed and inscribed figures are to one another in the duplicate ratio of their sides, and*

(2) *the figures themselves [i.e. their volumes] are in the triplicate ratio of their sides.*

(1) Let AA', aa' be in the same straight line, and let $MmOm'M'$ be a diameter at right angles to them.

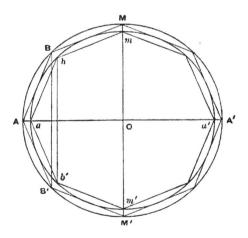

Join BB', CC',... and bb', cc',... which will all be parallel to one another and MM'.

Suppose R, S to be circles such that

$R = $ (surface of circumscribed solid),

$S = $ (surface of inscribed solid).

ON THE SPHERE AND CYLINDER I. 39

Then (radius of $R)^2 = AB(BB' + CC' + ...)$ [Prop. 29]

(radius of $S)^2 = ab(bb' + cc' + ...).$ [Prop. 24]

And, since the polygons are similar, the rectangles in these two equations are similar, and are therefore in the ratio of
$$AB^2 : ab^2.$$
Hence

(surface of circumscribed solid) : (surface of inscribed solid)
$$= AB^2 : ab^2.$$

(2) Take a cone V whose base is the circle R and whose height is equal to Oa, and a cone W whose base is the circle S and whose height is equal to the perpendicular from O on ab, which we will call p.

Then V, W are respectively equal to the volumes of the circumscribed and inscribed figures. [Props. 31, 26]

Now, since the polygons are similar,

$AB : ab = Oa : p$

 = (height of cone V) : (height of cone W);

and, as shown above, the bases of the cones (the circles R, S) are in the ratio of AB^2 to ab^2.

Therefore $\qquad V : W = AB^3 : ab^3.$

Proposition 33.

The surface of any sphere is equal to four times the greatest circle in it.

Let C be a circle equal to four times the great circle.

Then, if C is not equal to the surface of the sphere, it must either be less or greater.

I. Suppose C less than the surface of the sphere.

It is then possible to find two lines β, γ, of which β is the greater, such that

$\beta : \gamma <$ (surface of sphere) : C. [Prop. 2]

Take such lines, and let δ be a mean proportional between them.

Suppose similar regular polygons with $4n$ sides circumscribed about and inscribed in a great circle such that the ratio of their sides is less than the ratio $\beta : \delta$. [Prop. 3]

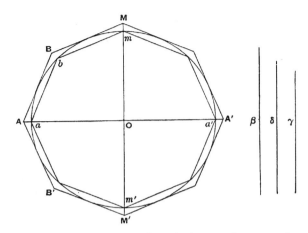

Let the polygons with the circle revolve together about a diameter common to all, describing solids of revolution as before.

Then (surface of outer solid) : (surface of inner solid)

$= $ (side of outer)^2 : (side of inner)^2 [Prop. 32]

$< \beta^2 : \delta^2$, or $\beta : \gamma$

$< $ (surface of sphere) : C, *a fortiori*.

But this is impossible, since the surface of the circumscribed solid is greater than that of the sphere [Prop. 28], while the surface of the inscribed solid is less than C [Prop. 25].

Therefore C is not less than the surface of the sphere.

II. Suppose C greater than the surface of the sphere.

Take lines β, γ, of which β is the greater, such that

$\beta : \gamma < C :$ (surface of sphere).

Circumscribe and inscribe to the great circle similar regular polygons, as before, such that their sides are in a ratio less than that of β to δ, and suppose solids of revolution generated in the usual manner.

Then, in this case,

(surface of circumscribed solid) : (surface of inscribed solid)

$< C$: (surface of sphere).

But this is impossible, because the surface of the circumscribed solid is greater than C [Prop. 30], while the surface of the inscribed solid is less than that of the sphere [Prop. 23].

Thus C is not greater than the surface of the sphere.

Therefore, since it is neither greater nor less, C is equal to the surface of the sphere.

Proposition 34.

Any sphere is equal to four times the cone which has its base equal to the greatest circle in the sphere and its height equal to the radius of the sphere.

Let the sphere be that of which $ama'm'$ is a great circle.

If now the sphere is not equal to four times the cone described, it is either greater or less.

I. If possible, let the sphere be greater than four times the cone.

Suppose V to be a cone whose base is equal to four times the great circle and whose height is equal to the radius of the sphere.

Then, by hypothesis, the sphere is greater than V; and two lines β, γ can be found (of which β is the greater) such that

$\beta : \gamma <$ (volume of sphere) : V.

Between β and γ place two arithmetic means δ, ϵ.

As before, let similar regular polygons with sides $4n$ in number be circumscribed about and inscribed in the great circle, such that their sides are in a ratio less than $\beta : \delta$.

Imagine the diameter aa' of the circle to be in the same straight line with a diameter of both polygons, and imagine the latter to revolve with the circle about aa', describing the surfaces of two solids of revolution. The volumes of these solids are therefore in the triplicate ratio of their sides.   [Prop. 32]

Thus (vol. of outer solid) : (vol. of inscribed solid)

$< \beta^3 : \delta^3$, by hypothesis,

$< \beta : \gamma$, a fortiori (since $\beta : \gamma > \beta^3 : \delta^3$)*,

$<$ (volume of sphere) : V, a fortiori.

But this is impossible, since the volume of the circumscribed

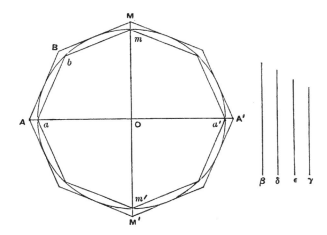

* That $\beta : \gamma > \beta^3 : \delta^3$ is assumed by Archimedes.   Eutocius proves the property in his commentary as follows.

Take x such that $\qquad \beta : \delta = \delta : x.$
Thus $\qquad \beta - \delta : \beta = \delta - x : \delta$
and, since $\beta > \delta$, $\beta - \delta > \delta - x$.
But, by hypothesis, $\qquad \beta - \delta = \delta - \epsilon.$
Therefore $\qquad \delta - \epsilon > \delta - x,$
or $\qquad x > \epsilon.$
Again, suppose $\qquad \delta : x = x : y,$
and, as before, we have $\qquad \delta - x > x - y,$
so that, a fortiori, $\qquad \delta - \epsilon > x - y.$
Therefore $\qquad \epsilon - \gamma > x - y\ ;$
and, since $x > \epsilon$, $y > \gamma$.

Now, by hypothesis, β, δ, x, y are in continued proportion;
therefore $\qquad \beta^3 : \delta^3 = \beta : y$
$\qquad\qquad\qquad\ \ < \beta : \gamma.$

solid is greater than that of the sphere [Prop. 28], while the volume of the inscribed solid is less than V [Prop. 27].

Hence the sphere is not greater than V, or four times the cone described in the enunciation.

II. If possible, let the sphere be less than V.

In this case we take β, γ (β being the greater) such that

$$\beta : \gamma < V : \text{(volume of sphere)}.$$

The rest of the construction and proof proceeding as before, we have finally

(volume of outer solid) : (volume of inscribed solid)

$$< V : \text{(volume of sphere)}.$$

But this is impossible, because the volume of the outer solid is greater than V [Prop. 31, Cor.], and the volume of the inscribed solid is less than the volume of the sphere.

Hence the sphere is not less than V.

Since then the sphere is neither less nor greater than V, it is equal to V, or to four times the cone described in the enunciation.

COR. From what has been proved it follows that *every cylinder whose base is the greatest circle in a sphere and whose height is equal to the diameter of the sphere is $\frac{3}{2}$ of the sphere, and its surface together with its bases is $\frac{3}{2}$ of the surface of the sphere.*

For the cylinder is three times the cone with the same base and height [Eucl. XII. 10], i.e. six times the cone with the same base and with height equal to the radius of the sphere.

But the sphere is four times the latter cone [Prop. 34]. Therefore the cylinder is $\frac{3}{2}$ of the sphere.

Again, the surface of a cylinder (excluding the bases) is equal to a circle whose radius is a mean proportional between the height of the cylinder and the diameter of its base [Prop. 13].

In this case the height is equal to the diameter of the base and therefore the circle is that whose radius is the diameter of the sphere, or a circle equal to four times the great circle of the sphere.

Therefore the surface of the cylinder with the bases is equal to six times the great circle.

And the surface of the sphere is four times the great circle [Prop. 33]; whence

(surface of cylinder with bases) = $\frac{3}{2}$. (surface of sphere).

Proposition 35.

If in a segment of a circle LAL' (where A is the middle point of the arc) a polygon LK...A...K'L' be inscribed of which LL' is one side, while the other sides are 2n in number and all equal, and if the polygon revolve with the segment about the diameter AM, generating a solid figure inscribed in a segment of a sphere, then the surface of the inscribed solid is equal to a circle the square on whose radius is equal to the rectangle

$$AB\left(BB' + CC' + \ldots + KK' + \frac{LL'}{2}\right).$$

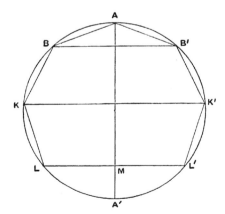

The surface of the inscribed figure is made up of portions of surfaces of cones.

If we take these successively, the surface of the cone BAB' is equal to a circle whose radius is

$$\sqrt{AB \cdot \tfrac{1}{2}BB'}.\qquad \text{[Prop. 14]}$$

The surface of the frustum of a cone $BCC'B'$ is equal to a circle whose radius is

$$\sqrt{AB \cdot \frac{BB' + CC'}{2}};\qquad \text{[Prop. 16]}$$

and so on.

Proceeding in this way and adding, we find, since circles are to one another as the squares of their radii, that the surface of the inscribed figure is equal to a circle whose radius is

$$\sqrt{AB\left(BB' + CC' + \ldots + KK' + \frac{LL'}{2}\right)}.$$

Proposition 36.

The surface of the figure inscribed as before in the segment of a sphere is less than that of the segment of the sphere.

This is clear, because the circular base of the segment is a common boundary of each of two surfaces, of which one, the segment, includes the other, the solid, while both are concave in the same direction [*Assumptions*, 4].

Proposition 37.

The surface of the solid figure inscribed in the segment of the sphere by the revolution of $LK\ldots A\ldots K'L'$ about AM is less than a circle with radius equal to AL.

Let the diameter AM meet the circle of which LAL' is a segment again in A'. Join $A'B$.

As in Prop. 35, the surface of the inscribed solid is equal to a circle the square on whose radius is

$$AB(BB' + CC' + \ldots + KK' + LM).$$

But this rectangle $= A'B \cdot AM$ [Prop. 22]
$< A'A \cdot AM$
$< AL^2$.

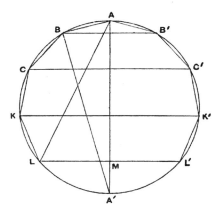

Hence the surface of the inscribed solid is less than the circle whose radius is AL.

Proposition 38.

The solid figure described as before in a segment of a sphere less than a hemisphere, together with the cone whose base is the base of the segment and whose apex is the centre of the sphere, is equal to a cone whose base is equal to the surface of the inscribed solid and whose height is equal to the perpendicular from the centre of the sphere on any side of the polygon.

Let O be the centre of the sphere, and p the length of the perpendicular from O on AB.

Suppose cones described with O as apex, and with the circles on BB', CC', ... as diameters as bases.

Then the rhombus $OBAB'$ is equal to a cone whose base is equal to the surface of the cone BAB', and whose height is p.

[Prop. 18]

Again, if CB, $C'B'$ meet in T, the solid described by the triangle BOC as the polygon revolves about AO is the difference between the rhombi $OCTC'$ and $OBTB'$, and is therefore equal to a cone whose base is equal to the surface of the frustum $BCC'B'$ and whose height is p. [Prop. 20]

Similarly for the part of the solid described by the triangle COD as the polygon revolves; and so on.

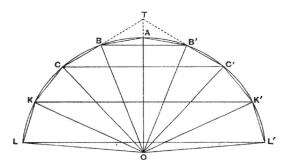

Hence, by addition, the solid figure inscribed in the segment together with the cone OLL' is equal to a cone whose base is the surface of the inscribed solid and whose height is p.

COR. *The cone whose base is a circle with radius equal to AL and whose height is equal to the radius of the sphere is greater than the sum of the inscribed solid and the cone OLL'.*

For, by the proposition, the inscribed solid together with the cone OLL' is equal to a cone with base equal to the surface of the solid and with height p.

This latter cone is less than a cone with height equal to OA and with base equal to the circle whose radius is AL, because the height p is less than OA, while the surface of the solid is less than a circle with radius AL. [Prop. 37]

Proposition 39.

Let lal' be a segment of a great circle of a sphere, being less than a semicircle. Let O be the centre of the sphere, and join Ol, Ol'. Suppose a polygon circumscribed about the sector $Olal'$ such that its sides, excluding the two radii, are $2n$ in number and all equal, as $LK, \ldots BA, AB', \ldots K'L'$; and let OA be that radius of the great circle which bisects the segment lal'.

The circle circumscribing the polygon will then have the same centre O as the given great circle.

Now suppose the polygon and the two circles to revolve together about OA. The two circles will describe spheres, the

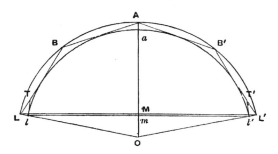

angular points except A will describe circles on the outer sphere, with diameters BB' etc., the points of contact of the sides with the inner segment will describe circles on the inner sphere, the sides themselves will describe the surfaces of cones or frusta of cones, and the whole figure circumscribed to the segment of the inner sphere by the revolution of the equal sides of the polygon will have for its base the circle on LL' as diameter.

The surface of the solid figure so circumscribed about the sector of the sphere [excluding its base] will be greater than that of the segment of the sphere whose base is the circle on ll' as diameter.

For draw the tangents lT, $l'T'$ to the inner segment at l, l'. These with the sides of the polygon will describe by their revolution a solid whose surface is greater than that of the segment [*Assumptions*, 4].

But the surface described by the revolution of lT is less than that described by the revolution of LT, since the angle TlL is a right angle, and therefore $LT > lT$.

Hence, *a fortiori*, the surface described by $LK\ldots A\ldots K'L'$ is greater than that of the segment.

ON THE SPHERE AND CYLINDER I. 49

Cor. *The surface of the figure so described about the sector of the sphere is equal to a circle the square on whose radius is equal to the rectangle*

$$AB(BB' + CC' + \ldots + KK' + \tfrac{1}{2}LL').$$

For the circumscribed figure is inscribed in the outer sphere, and the proof of Prop. 35 therefore applies.

Proposition 40.

The surface of the figure circumscribed to the sector as before is greater than a circle whose radius is equal to al.

Let the diameter AaO meet the great circle and the circle circumscribing the revolving polygon again in a', A'. Join $A'B$, and let ON be drawn to N, the point of contact of AB with the inner circle.

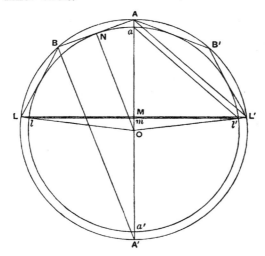

Now, by Prop. 39, Cor., the surface of the solid figure circumscribed to the sector $OlAl'$ is equal to a circle the square on whose radius is equal to the rectangle

$$AB\left(BB' + CC' + \ldots + KK' + \frac{LL'}{2}\right).$$

But this rectangle is equal to $A'B \cdot AM$ [as in Prop. 22].

Next, since AL', al' are parallel, the triangles AML', aml' are similar. And $AL' > al'$; therefore $AM > am$.

Also $\qquad A'B = 2ON = aa'$.

Therefore $\qquad A'B \cdot AM > am \cdot aa'$

$\qquad\qquad\qquad > al'^2$.

Hence the surface of the solid figure circumscribed to the sector is greater than a circle whose radius is equal to al', or al.

Cor. 1. *The volume of the figure circumscribed about the sector together with the cone whose apex is O and base the circle on LL' as diameter, is equal to the volume of a cone whose base is equal to the surface of the circumscribed figure and whose height is ON.*

For the figure is inscribed in the outer sphere which has the same centre as the inner. Hence the proof of Prop. 38 applies.

Cor. 2. *The volume of the circumscribed figure with the cone OLL' is greater than the cone whose base is a circle with radius equal to al and whose height is equal to the radius (Oa) of the inner sphere.*

For the volume of the figure with the cone OLL' is equal to a cone whose base is equal to the surface of the figure and whose height is equal to ON.

And the surface of the figure is greater than a circle with radius equal to al [Prop. 40], while the heights Oa, ON are equal.

Proposition 41.

Let lal' be a segment of a great circle of a sphere which is less than a semicircle.

Suppose a polygon inscribed in the sector $Olal'$ such that the sides $lk, \ldots ba, ab', \ldots k'l'$ are $2n$ in number and all equal. Let a similar polygon be circumscribed about the sector so that its sides are parallel to those of the first polygon; and draw the circle circumscribing the outer polygon.

Now let the polygons and circles revolve together about OaA, the radius bisecting the segment lal'.

Then (1) *the surfaces of the outer and inner solids of revolution so described are in the ratio of AB^2 to ab^2, and* (2) *their volumes together with the corresponding cones with the same base and with apex O in each case are as AB^3 to ab^3.*

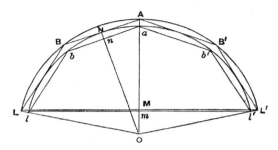

(1) For the surfaces are equal to circles the squares on whose radii are equal respectively to

$$AB\left(BB' + CC' + \ldots + KK' + \frac{LL'}{2}\right),$$

[Prop. 39, Cor.]

and $$ab\left(bb' + cc' + \ldots + kk' + \frac{ll'}{2}\right).$$ [Prop. 35]

But these rectangles are in the ratio of AB^2 to ab^2. Therefore so are the surfaces.

(2) Let OnN be drawn perpendicular to ab and AB; and suppose the circles which are equal to the surfaces of the outer and inner solids of revolution to be denoted by S, s respectively.

Now the volume of the circumscribed solid together with the cone OLL' is equal to a cone whose base is S and whose height is ON [Prop. 40, Cor. 1].

And the volume of the inscribed figure with the cone Oll' is equal to a cone with base s and height On [Prop. 38].

But $S : s = AB^2 : ab^2$,

and $ON : On = AB : ab$.

Therefore the volume of the circumscribed solid together with the cone OLL' is to the volume of the inscribed solid together with the cone Oll' as AB^3 is to ab^3 [*Lemma* 5].

Proposition 42.

If lal' be a segment of a sphere less than a hemisphere and Oa the radius perpendicular to the base of the segment, the surface of the segment is equal to a circle whose radius is equal to al.

Let R be a circle whose radius is equal to al. Then the surface of the segment, which we will call S, must, if it be not equal to R, be either greater or less than R.

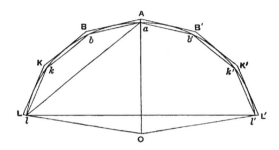

I. Suppose, if possible, $S > R$.

Let lal' be a segment of a great circle which is less than a semicircle. Join Ol, Ol', and let similar polygons with $2n$ equal sides be circumscribed and inscribed to the sector, as in the previous propositions, but such that

(circumscribed polygon) : (inscribed polygon) $< S : R$.

[Prop. 6]

Let the polygons now revolve with the segment about OaA, generating solids of revolution circumscribed and inscribed to the segment of the sphere.

Then

(surface of outer solid) : (surface of inner solid)

$= AB^2 : ab^2$ [Prop. 41]

$=$ (circumscribed polygon) : (inscribed polygon)

$< S : R$, by hypothesis.

But the surface of the outer solid is greater than S [Prop. 39].

Therefore the surface of the inner solid is greater than R; which is impossible, by Prop. 37.

II. Suppose, if possible, $S < R$.

In this case we circumscribe and inscribe polygons such that their ratio is less than $R : S$; and we arrive at the result that

(surface of outer solid) : (surface of inner solid)
$$< R : S.$$

But the surface of the outer solid is greater than R [Prop. 40]. Therefore the surface of the inner solid is greater than S: which is impossible [Prop. 36].

Hence, since S is neither greater nor less than R,
$$S = R.$$

Proposition 43.

Even if the segment of the sphere is greater than a hemisphere, its surface is still equal to a circle whose radius is equal to al.

For let $lal'a'$ be a great circle of the sphere, aa' being the diameter perpendicular to ll'; and let $la'l'$ be a segment less than a semicircle.

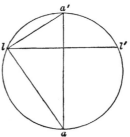

Then, by Prop. 42, the surface of the segment $la'l'$ of the sphere is equal to a circle with radius equal to $a'l$.

Also the surface of the whole sphere is equal to a circle with radius equal to aa' [Prop. 33].

But $aa'^2 - a'l^2 = al^2$, and circles are to one another as the squares on their radii.

Therefore the surface of the segment lal', being the difference between the surfaces of the sphere and of $la'l'$, is equal to a circle with radius equal to al.

Proposition 44.

The volume of any sector of a sphere is equal to a cone whose base is equal to the surface of the segment of the sphere included in the sector, and whose height is equal to the radius of the sphere.

Let R be a cone whose base is equal to the surface of the segment lal' of a sphere and whose height is equal to the radius of the sphere; and let S be the volume of the sector $Olal'$.

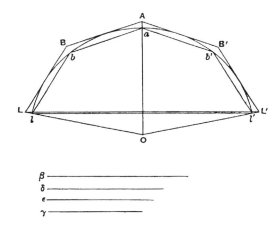

Then, if S is not equal to R, it must be either greater or less.

I. Suppose, if possible, that $S > R$.

Find two straight lines β, γ, of which β is the greater, such that
$$\beta : \gamma < S : R;$$
and let δ, ϵ be two arithmetic means between β, γ.

Let lal' be a segment of a great circle of the sphere. Join Ol, Ol', and let similar polygons with $2n$ equal sides be circumscribed and inscribed to the sector of the circle as before, but such that their sides are in a ratio less than $\beta : \delta$. [Prop. 4].

Then let the two polygons revolve with the segment about OaA, generating two solids of revolution.

Denoting the volumes of these solids by V, v respectively, we have

$(V + \text{cone } OLL') : (v + \text{cone } Oll') = AB^3 : ab^3$ [Prop. 41]
$\qquad\qquad\qquad\qquad\qquad < \beta^3 : \delta^3$
$\qquad\qquad\qquad\qquad\qquad < \beta : \gamma$, *a fortiori**,
$\qquad\qquad\qquad\qquad\qquad < S : R$, by hypothesis.

Now $\qquad (V + \text{cone } OLL') > S.$
Therefore also $\qquad (v + \text{cone } Oll') > R.$

But this is impossible, by Prop. 38, Cor. combined with Props. 42, 43.

Hence $\qquad\qquad S \not> R.$

II. Suppose, if possible, that $S < R$.

In this case we take β, γ such that

$$\beta : \gamma < R : S,$$

and the rest of the construction proceeds as before.

We thus obtain the relation

$\qquad (V + \text{cone } OLL') : (v + \text{cone } Oll') < R : S.$

Now $\qquad (v + \text{cone } Oll') < S.$
Therefore $\qquad (V + \text{cone } OLL') < R;$

which is impossible, by Prop. 40, Cor. 2 combined with Props. 42, 43.

Since then S is neither greater nor less than R,

$$S = R.$$

---

* Cf. note on Prop. 34, p. 42..

ON THE SPHERE AND CYLINDER.

BOOK II.

"ARCHIMEDES to Dositheus greeting.

On a former occasion you asked me to write out the proofs of the problems the enunciations of which I had myself sent to Conon. In point of fact they depend for the most part on the theorems of which I have already sent you the demonstrations, namely (1) that the surface of any sphere is four times the greatest circle in the sphere, (2) that the surface of any segment of a sphere is equal to a circle whose radius is equal to the straight line drawn from the vertex of the segment to the circumference of its base, (3) that the cylinder whose base is the greatest circle in any sphere and whose height is equal to the diameter of the sphere is itself in magnitude half as large again as the sphere, while its surface [including the two bases] is half as large again as the surface of the sphere, and (4) that any solid sector is equal to a cone whose base is the circle which is equal to the surface of the segment of the sphere included in the sector, and whose height is equal to the radius of the sphere. Such then of the theorems and problems as depend on these theorems I have written out in the book which I send herewith; those which are discovered by means of a different sort of investigation, those namely which relate to spirals and the conoids, I will endeavour to send you soon.

The first of the problems was as follows: *Given a sphere, to find a plane area equal to the surface of the sphere.*

The solution of this is obvious from the theorems aforesaid. For four times the greatest circle in the sphere is both a plane area and equal to the surface of the sphere.

The second problem was the following."

Proposition 1. (Problem.)

Given a cone or a cylinder, to find a sphere equal to the cone or to the cylinder.

If V be the given cone or cylinder, we can make a cylinder equal to $\frac{3}{2}V$. Let this cylinder be the cylinder whose base is the circle on AB as diameter and whose height is OD.

Now, if we could make another cylinder, equal to the cylinder (OD) but such that its height is equal to the diameter of its base, the problem would be solved, because this latter cylinder would be equal to $\frac{3}{2}V$, and the sphere whose diameter is equal to the height (or to the diameter of the base) of the same cylinder would then be the sphere required [I. 34, Cor.].

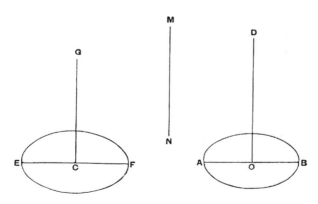

Suppose the problem solved, and let the cylinder (CG) be equal to the cylinder (OD), while EF, the diameter of the base, is equal to the height CG.

Then, since in equal cylinders the heights and bases are reciprocally proportional,

$$AB^2 : EF^2 = CG : OD$$
$$= EF : OD \dots\dots\dots\dots\dots(1).$$

Suppose MN to be such a line that

$$EF^2 = AB \cdot MN \dots\dots\dots\dots\dots\dots(2).$$

Hence $\quad AB : EF = EF : MN$,

and, combining (1) and (2), we have

$$AB : MN = EF : OD,$$

or $\quad\quad\quad AB : EF = MN : OD.$

Therefore $\quad AB : EF = EF : MN = MN : OD$,

and EF, MN *are two mean proportionals between AB, OD*.

The synthesis of the problem is therefore as follows. Take two mean proportionals EF, MN between AB and OD, and describe a cylinder whose base is a circle on EF as diameter and whose height CG is equal to EF.

Then, since

$$AB : EF = EF : MN = MN : OD,$$
$$EF^2 = AB \cdot MN,$$

and therefore $\quad AB^2 : EF^2 = AB : MN$
$$= EF : OD$$
$$= CG : OD;$$

whence the bases of the two cylinders (OD), (CG) are reciprocally proportional to their heights.

Therefore the cylinders are equal, and it follows that

$$\text{cylinder } (CG) = \tfrac{3}{2} V.$$

The sphere on EF as diameter is therefore the sphere required, being equal to V.

Proposition 2.

If BAB' be a segment of a sphere, BB' a diameter of the base of the segment, and O the centre of the sphere, and if AA' be the diameter of the sphere bisecting BB' in M, then the volume of the segment is equal to that of a cone whose base is the same as that of the segment and whose height is h, where

$$h : AM = OA' + A'M : A'M.$$

Measure MH along MA equal to h, and MH' along MA' equal to h', where

$$h' : A'M = OA + AM : AM.$$

Suppose the three cones constructed which have O, H H' for their apices and the base (BB') of the segment for their common base. Join AB, $A'B$.

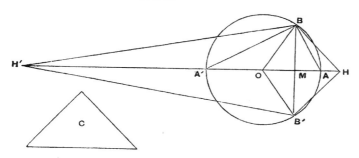

Let C be a cone whose base is equal to the surface of the segment BAB' of the sphere, i.e. to a circle with radius equal to AB [I. 42], and whose height is equal to OA.

Then the cone C is equal to the solid sector $OBAB'$ [I. 44].

Now, since $HM : MA = OA' + A'M : A'M$,

dividendo, $\qquad HA : AM = OA : A'M$,

and, alternately, $HA : AO = AM : MA'$,

so that

$$HO : OA = AA' : A'M$$
$$= AB^2 : BM^2$$
$$= \text{(base of cone } C\text{)} : \text{(circle on } BB' \text{ as diameter)}.$$

But *OA* is equal to the height of the cone C; therefore, since cones are equal if their bases and heights are reciprocally proportional, it follows that the cone C (or the solid sector $OBAB'$) is equal to a cone whose base is the circle on BB' as diameter and whose height is equal to OH.

And this latter cone is equal to the sum of two others having the same base and with heights OM, MH, i.e. to the solid rhombus $OBHB'$.

Hence the sector $OBAB'$ is equal to the rhombus $OBHB'$.

Taking away the common part, the cone OBB',

the segment BAB' = the cone HBB'.

Similarly, by the same method, we can prove that

the segment $BA'B'$ = the cone $H'BB'$.

Alternative proof of the latter property.

Suppose D to be a cone whose base is equal to the surface of the whole sphere and whose height is equal to OA.

Thus D is equal to the volume of the sphere. [I. 33, 34]

Now, since $OA' + A'M : A'M = HM : MA$,

dividendo and *alternando*, as before,

$$OA : AH = A'M : MA.$$

Again, since $H'M : MA' = OA + AM : AM$,

$$H'A' : OA = A'M : MA$$
$$= OA : AH, \text{ from above.}$$

Componendo, $\quad H'O : OA = OH : HA$ ....(1).

Alternately, $\quad H'O : OH = OA : AH$ ....(2),

and, *componendo,* $HH' : HO = OH : HA$,

$\qquad\qquad = H'O : OA$, from (1),

whence $\qquad HH' . OA = H'O . OH$ ....(3).

Next, since $\quad H'O : OH = OA : AH$, by (2),

$\qquad\qquad = A'M : MA$,

$(H'O + OH)^2 : H'O . OH = (A'M + MA)^2 : A'M . MA$,

whence, by means of (3),
$$HH'^2 : HH' . OA = AA'^2 : A'M . MA,$$
or $\qquad HH' : OA = AA'^2 : BM^2.$

Now the cone D, which is equal to the sphere, has for its base a circle whose radius is equal to AA', and for its height a line equal to OA.

Hence this cone D is equal to a cone whose base is the circle on BB' as diameter and whose height is equal to HH';

therefore the cone D = the rhombus $HBH'B'$,
or the rhombus $HBH'B'$ = the sphere.
But the segment BAB' = the cone HBB';
therefore the remaining segment $BA'B'$ = the cone $H'BB'$.

COR. *The segment BAB' is to a cone with the same base and equal height in the ratio of $OA' + A'M$ to $A'M$.*

Proposition 3. (Problem.)

To cut a given sphere by a plane so that the surfaces of the segments may have to one another a given ratio.

Suppose the problem solved. Let AA' be a diameter of a great circle of the sphere, and suppose that a plane perpendicular to AA' cuts the plane of the great circle in the straight

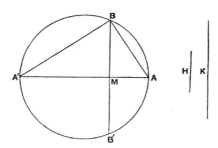

line BB', and AA' in M, and that it divides the sphere so that the surface of the segment BAB' has to the surface of the segment $BA'B'$ the given ratio.

Now these surfaces are respectively equal to circles with radii equal to AB, $A'B$ [I. 42, 43].

Hence the ratio $AB^2 : A'B^2$ is equal to the given ratio, i.e. AM is to MA' in the given ratio.

Accordingly the synthesis proceeds as follows.

If $H : K$ be the given ratio, divide AA' in M so that
$$AM : MA' = H : K.$$
Then $AM : MA' = AB^2 : A'B^2$
= (circle with radius AB) : (circle with radius $A'B$)
= (surface of segment BAB') : (surface of segment $BA'B'$).

Thus the ratio of the surfaces of the segments is equal to the ratio $H : K$.

Proposition 4. (Problem.)

To cut a given sphere by a plane so that the volumes of the segments are to one another in a given ratio.

Suppose the problem solved, and let the required plane cut the great circle ABA' at right angles in the line BB'. Let AA' be that diameter of the great circle which bisects BB' at right angles (in M), and let O be the centre of the sphere.

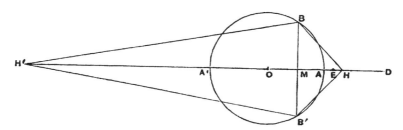

Take H on OA produced, and H' on OA' produced, such that
$$OA' + A'M : A'M = HM : MA, \quad\quad\quad (1),$$
and
$$OA + AM : AM = H'M : MA' \quad\quad\quad (2).$$

Join BH, $B'H$, BH', $B'H'$.

ON THE SPHERE AND CYLINDER II. 63

Then the cones HBB', $H'BB'$ are respectively equal to the segments BAB', $BA'B'$ of the sphere [Prop. 2].

Hence the ratio of the cones, and therefore of their altitudes, is given, i.e.

$$HM : H'M = \text{the given ratio} \dots \dots \dots (3).$$

We have now three equations (1), (2), (3), in which there appear three as yet undetermined points M, H, H'; and it is first necessary to find, by means of them, another equation in which only one of these points (M) appears, i.e. we have, so to speak, to *eliminate* H, H'.

Now, from (3), it is clear that $HH' : H'M$ is also a given ratio; and Archimedes' method of elimination is, *first*, to find values for each of the ratios $A'H' : H'M$ and $HH' : H'A'$ which are alike independent of H, H', and then, *secondly*, to equate the ratio compounded of these two ratios to the known value of the ratio $HH' : H'M$.

(a) To find such a value for $A'H' : H'M$.

It is at once clear from equation (2) above that

$$A'H' : H'M = OA : OA + AM \dots \dots \dots (4).$$

(b) To find such a value for $HH' : A'H'$.

From (1) we derive

$$A'M : MA = OA' + A'M : HM$$
$$= OA' : AH \dots \dots \dots \dots \dots (5);$$

and, from (2), $\quad A'M : MA = H'M : OA + AM$
$$= A'H' : OA \dots \dots \dots \dots \dots (6).$$

Thus $\quad HA : AO = OA' : A'H'$,

whence $\quad OH : OA' = OH' : A'H'$,

or $\quad OH : OH' = OA' : A'H'$.

It follows that

$$HH' : OH' = OH' : A'H',$$

or $\quad HH' . H'A' = OH'^2$.

Therefore $\quad HH' : H'A' = OH'^2 : H'A'^2$
$$= AA'^2 : A'M^2, \text{ by means of (6)}$$

(c) To express the ratios $A'H' : H'M$ and $HH' : H'M$ more simply we make the following construction. Produce OA to D so that $OA = AD$. (D will lie beyond H, for $A'M > MA$, and therefore, by (5), $OA > AH$.)

Then
$$A'H' : H'M = OA : OA + AM$$
$$= AD : DM \ldots\ldots\ldots\ldots\ldots\ldots(7).$$

Now divide AD at E so that
$$HH' : H'M = AD : DE \ldots\ldots\ldots\ldots\ldots(8).$$

Thus, using equations (8), (7) and the value of $HH' : H'A'$ above found, we have
$$AD : DE = HH' : H'M$$
$$= (HH' : H'A') . (A'H' : H'M)$$
$$= (AA'^2 : A'M^2) . (AD : DM).$$

But $\quad AD : DE = (DM : DE) . (AD : DM).$

Therefore $\quad MD : DE = AA'^2 : A'M^2 \ldots\ldots\ldots\ldots\ldots(9).$

And D is given, since $AD = OA$. Also $AD : DE$ (being equal to $HH' : H'M$) is a given ratio. Therefore DE is given.

Hence the problem reduces itself to the problem of dividing $A'D$ into two parts at M so that

$$MD : \text{(a given length)} = \text{(a given area)} : A'M^2.$$

Archimedes adds: "If the problem is propounded in this general form, it requires a διορισμός [i.e. it is necessary to investigate the limits of possibility], but, if there be added the conditions subsisting in the present case, it does not require a διορισμός."

In the present case the problem is:

Given a straight line $A'A$ produced to D so that $A'A = 2AD$, and given a point E on AD, to cut AA' in a point M so that

$$AA'^2 : A'M^2 = MD : DE.$$

"And the analysis and synthesis of both problems will be given at the end*."

The synthesis of the main problem will be as follows. Let $R : S$ be the given ratio, R being less than S. AA' being a

---

* See the note following this proposition.

diameter of a great circle, and O the centre, produce OA to D so that $OA = AD$, and divide AD in E so that
$$AE : ED = R : S.$$
Then cut AA' in M so that
$$MD : DE = AA'^2 : A'M^2.$$
Through M erect a plane perpendicular to AA'; this plane will then divide the sphere into segments which will be to one another as R to S.

Take H on $A'A$ produced, and H' on AA' produced, so that
$$OA' + A'M : A'M = HM : MA, \dots\dots\dots\dots(1),$$
$$OA + AM : AM = H'M : MA'\dots\dots\dots\dots(2).$$
We have then to show that
$$HM : MH' = R : S, \text{ or } AE : ED.$$

(a) We first find the value of $HH' : H'A'$ as follows.

As was shown in the analysis (b),
$$HH' \cdot H'A' = OH'^2,$$
or
$$HH' : H'A' = OH'^2 : H'A'^2$$
$$= AA'^2 : A'M^2$$
$$= MD : DE, \text{ by construction.}$$

(β) Next we have
$$H'A' : H'M = OA : OA + AM$$
$$= AD : DM.$$
Therefore
$$HH' : H'M = (HH' : H'A') \cdot (H'A' : H'M)$$
$$= (MD : DE) \cdot (AD : DM)$$
$$= AD : DE,$$
whence
$$HM : MH' = AE : ED$$
$$= R : S. \qquad \text{Q. E. D.}$$

Note. The solution of the subsidiary problem to which the original problem of Prop. 4 is reduced, and of which Archimedes promises a discussion, is given in a highly interesting and important note by Eutocius, who introduces the subject with the following explanation.

"He [Archimedes] promised to give a solution of this problem at the end, but we do not find the promise kept in any of the copies. Hence we find that Dionysodorus too failed to light upon the promised discussion and, being unable to grapple with the omitted lemma, approached the original problem in a different way, which I shall describe later. Diocles also expressed in his work περὶ πυρίων the opinion that Archimedes made the promise but did not perform it, and tried to supply the omission himself. His attempt I shall also give in its order. It will however be seen to have no relation to the omitted discussion but to give, like Dionysodorus, a construction arrived at by a different method of proof. On the other hand, as the result of unremitting and extensive research, I found in a certain old book some theorems discussed which, although the reverse of clear owing to errors and in many ways faulty as regards the figures, nevertheless gave the substance of what I sought, and moreover to some extent kept to the Doric dialect affected by Archimedes, while they retained the names familiar in old usage, the parabola being called a section of a right-angled cone, and the hyperbola a section of an obtuse-angled cone; whence I was led to consider whether these theorems might not in fact be what he promised he would give at the end. For this reason I paid them the closer attention, and, after finding great difficulty with the actual text owing to the multitude of the mistakes above referred to, I made out the sense gradually and now proceed to set it out, as well as I can, in more familiar and clearer language. And first the theorem will be treated generally, in order that what Archimedes says about the limits of possibility may be made clear; after which there will follow the special application to the conditions stated in his analysis of the problem."

The investigation which follows may be thus reproduced. The general problem is:

Given two straight lines AB, AC and an area D, to divide AB at M so that

$$AM : AC = D : MB^2.$$

Analysis.

Suppose M found, and suppose AC placed at right angles to AB. Join CM and produce it. Draw EBN through B parallel to AC meeting CM in N, and through C draw CHE parallel to AB meeting EBN in E. Complete the parallelogram $CENF$, and through M draw PMH parallel to AC meeting FN in P.

Measure EL along EN so that

$CE \cdot EL$ (or $AB \cdot EL$) $= D$.

Then, by hypothesis,

$AM : AC = CE \cdot EL : MB^2$.

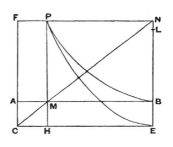

And

$AM : AC = CE : EN$,

by similar triangles,

$= CE \cdot EL : EL \cdot EN$.

It follows that $PN^2 = MB^2 = EL \cdot EN$.

Hence, if a parabola be described with vertex E, axis EN, and parameter equal to EL, it will pass through P; and it will be given in position, since EL is given.

Therefore P lies on a given parabola.

Next, since the rectangles FH, AE are equal,

$FP \cdot PH = AB \cdot BE$.

Hence, if a rectangular hyperbola be described with CE, CF as asymptotes and passing through B, it will pass through P. And the hyperbola is given in position.

Therefore P lies on a given hypérbola.

Thus P is determined as the intersection of the parabola and hyperbola. And since P is thus given, M is also given.

διορισμός.

Now, since $AM : AC = D : MB^2$,

$AM \cdot MB^2 = AC \cdot D$.

But $AC \cdot D$ is given, and *it will be proved later that the maximum value of $AM \cdot MB^2$ is that which it assumes when $BM = 2AM$.*

Hence *it is a necessary condition of the possibility of a solution that $AC \cdot D$ must not be greater than $\tfrac{1}{3}AB \cdot (\tfrac{2}{3}AB)^2$, or $\tfrac{4}{27}AB^2$.*

Synthesis.

If O be such a point on AB that $BO = 2AO$, we have seen that, in order that the solution may be possible,

$$AC \cdot D \not> AO \cdot OB^2.$$

Thus $AC \cdot D$ is either equal to, or less than, $AO \cdot OB^2$.

(1) If $AC \cdot D = AO \cdot OB^2$, then the point O itself solves the problem.

(2) Let $AC \cdot D$ be less than $AO \cdot OB^2$.

Place AC at right angles to AB. Join CO, and produce it to R. Draw EBR through B parallel to AC meeting CO in R, and through C draw CE parallel to AB meeting EBR in E. Complete the parallelogram $CERF$, and through O draw QOK parallel to AC meeting FR in Q and CE in K.

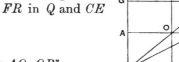

Then, since

$$AC \cdot D < AO \cdot OB^2,$$

measure RQ' along RQ so that

$$AC \cdot D = AO \cdot Q'R^2,$$

or $\qquad AO : AC = D : Q'R^2.$

Measure EL along ER so that

$$D = CE \cdot EL \text{ (or } AB \cdot EL).$$

Now, since $\quad AO : AC = D : Q'R^2$, by hypothesis,

$$= CE \cdot EL : Q'R^2,$$

and $\qquad AO : AC = CE : ER$, by similar triangles,

$$= CE \cdot EL : EL \cdot ER,$$

it follows that

$$Q'R^2 = EL \cdot ER.$$

Describe a parabola with vertex E, axis ER, and parameter equal to EL. This parabola will then pass through Q'.

Again, rect. FK = rect. AE,

or $FQ \cdot QK = AB \cdot BE$;

and, if we describe a rectangular hyperbola with asymptotes CE, CF and passing through B, it will also pass through Q.

Let the parabola and hyperbola intersect at P, and through P draw PMH parallel to AC meeting AB in M and CE in H, and GPN parallel to AB meeting CF in G and ER in N.

Then shall M be the required point of division.

Since $PG \cdot PH = AB \cdot BE$,

rect. GM = rect. ME,

and therefore CMN is a straight line.

Thus $AB \cdot BE = PG \cdot PH = AM \cdot EN$ ....(1).

Again, by the property of the parabola,

$PN^2 = EL \cdot EN$,

or $MB^2 = EL \cdot EN$ ....(2).

From (1) and (2)

$AM : EL = AB \cdot BE : MB^2$,

or $AM \cdot AB : AB \cdot EL = AB \cdot AC : MB^2$.

Alternately,

$AM \cdot AB : AB \cdot AC = AB \cdot EL : MB^2$,

or $AM : AC = D : MB^2$.

Proof of διορισμός.

It remains to be proved that, *if AB be divided at O so that $BO = 2AO$, then $AO \cdot OB^2$ is the maximum value of $AM \cdot MB^2$*,

or $AO \cdot OB^2 > AM \cdot MB^2$,

where M is any point on AB other than O.

Suppose that $AO : AC = CE \cdot EL' : OB^2$,
so that $AO \cdot OB^2 = CE \cdot EL' \cdot AC$.

Join CO, and produce it to N; draw EBN through B parallel to AC, and complete the parallelogram $CENF$.

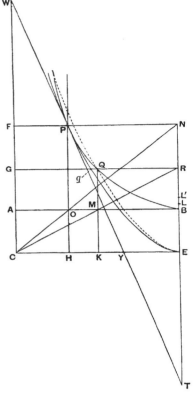

Through O draw POH parallel to AC meeting FN in P and CE in H.

With vertex E, axis EN, and parameter EL', describe a parabola. This will pass through P, as shown in the analysis above, and beyond P will meet the diameter CF of the parabola in some point.

Next draw a rectangular hyperbola with asymptotes CE, CF and passing through B. This hyperbola will also pass through P, as shown in the analysis.

Produce NE to T so that $TE = EN$. Join TP meeting CE in Y, and produce it to meet CF in W. Thus TP will touch the parabola at P.

Then, since $\qquad BO = 2AO$,
$\qquad\qquad\qquad\quad TP = 2PW$.
And $\qquad\qquad\quad TP = 2PY$.
Therefore $\qquad\quad PW = PY$.

Since, then, WY between the asymptotes is bisected at P, the point where it meets the hyperbola,

$\qquad\quad WY$ is a tangent to the hyperbola.

Hence the hyperbola and parabola, having a common tangent at P, touch one another at P.

Now take any point M on AB, and through M draw QMK parallel to AC meeting the hyperbola in Q and CE in K. Lastly, draw $GqQR$ through Q parallel to AB meeting CF in G, the parabola in q, and EN in R.

Then, since, by the property of the hyperbola, the rectangles GK, AE are equal, CMR is a straight line.

By the property of the parabola,
$$qR^2 = EL' \cdot ER,$$
so that $\quad QR^2 < EL' \cdot ER.$
Suppose $\quad QR^2 = EL \cdot ER,$
and we have $\quad AM : AC = CE : ER$
$$= CE \cdot EL : EL \cdot ER$$
$$= CE \cdot EL : QR^2$$
$$= CE \cdot EL : MB^2,$$
or $\quad AM \cdot MB^2 = CE \cdot EL \cdot AC.$
Therefore $\quad AM \cdot MB^2 < CE \cdot EL' \cdot AC$
$$< AO \cdot OB^2.$$

If $AC \cdot D < AO \cdot OB^2$, there are two solutions because there will be two points of intersection between the parabola and the hyperbola.

For, if we draw with vertex E and axis EN a parabola whose parameter is equal to EL, the parabola will pass through the point Q (see the last figure); and, since the parabola meets the diameter CF beyond Q, it must meet the hyperbola again (which has CF for its asymptote).

[If we put $AB = a$, $BM = x$, $AC = c$, and $D = b^2$, the proportion
$$AM : AC = D : MB^2$$
is seen to be equivalent to the equation
$$x^2(a - x) = b^2 c,$$
being a *cubic equation* with the term containing x omitted.

Now suppose EN, EC to be axes of coordinates, EN being the axis of y.

Then the parabola used in the above solution is the parabola
$$x^2 = \frac{b^2}{a} \cdot y,$$
and the rectangular hyperbola is
$$y(a-x) = ac.$$
Thus the solution of the cubic equation and the conditions under which there are no positive solutions, or one, or two positive solutions are obtained by the use of the two conics.]

[For the sake of completeness, and for their intrinsic interest, the solutions of the original problem in Prop. 4 given by Dionysodorus and Diocles are here appended.

Dionysodorus' solution.

Let AA' be a diameter of the given sphere. It is required to find a plane cutting AA' at right angles (in a point M, suppose) so that the segments into which the sphere is divided are in a given ratio, as $CD : DE$.

Produce $A'A$ to F so that $AF = OA$, where O is the centre of the sphere.

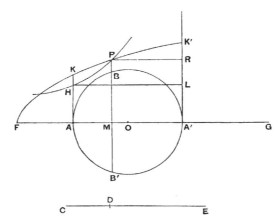

Draw AH perpendicular to AA' and of such length that
$$FA : AH = CE : ED,$$

and produce AH to K so that
$$AK^2 = FA \cdot AH \quad \ldots\ldots\ldots\ldots\ldots\ldots(\alpha).$$

With vertex F, axis FA, and parameter equal to AH describe a parabola. This will pass through K, by the equation (α).

Draw $A'K'$ parallel to AK and meeting the parabola in K'; and with $A'F$, $A'K'$ as asymptotes describe a rectangular hyperbola passing through H. This hyperbola will meet the parabola at some point, as P, between K and K'.

Draw PM perpendicular to AA' meeting the great circle in B, B', and from H, P draw HL, PR both parallel to AA' and meeting $A'K'$ in L, R respectively.

Then, by the property of the hyperbola,
$$PR \cdot PM = AH \cdot HL,$$
i.e. $$PM \cdot MA' = HA \cdot AA',$$
or $$PM : AH = AA' : A'M,$$
and $$PM^2 : AH^2 = AA'^2 : A'M^2.$$

Also, by the property of the parabola,
$$PM^2 = FM \cdot AH,$$
i.e. $$FM : PM = PM : AH,$$
or $$FM : AH = PM^2 : AH^2$$
$$= AA'^2 : A'M^2, \text{ from above.}$$

Thus, since circles are to one another as the squares of their radii, the cone whose base is the circle with $A'M$ as radius and whose height is equal to FM, and the cone whose base is the circle with AA' as radius and whose height is equal to AH, have their bases and heights reciprocally proportional.

Hence the cones are equal; i.e., if we denote the first cone by the symbol $c(A'M)$, FM, and so on,
$$c(A'M), FM = c(AA'), AH.$$
Now $c(AA'), FA : c(AA'), AH = FA : AH$
$$= CE : ED, \text{ by construction.}$$

Therefore

$$c(AA'), FA : c(A'M), FM = CE : ED \dots\dots(\beta).$$

But (1) $c(AA'), FA$ = the sphere.   [I. 34]

(2) $c(A'M), FM$ can be proved equal to the segment of the sphere whose vertex is A' and height $A'M$.

For take G on AA' produced such that

$$GM : MA' = FM : MA$$
$$= OA + AM : AM.$$

Then the cone GBB' is equal to the segment $A'BB'$ [Prop. 2].

And $FM : MG = AM : MA'$, by hypothesis,
$$= BM^2 : A'M^2.$$

Therefore

(circle with rad. BM) : (circle with rad. $A'M$)
$$= FM : MG,$$

so that $c(A'M), FM = c(BM), MG$

= the segment $A'BB'$.

We have therefore, from the equation (β) above,

(the sphere) : (segmt. $A'BB'$) = $CE : ED$,

whence (segmt. ABB') : (segmt. $A'BB'$) = $CD : DE$.

Diocles' solution.

Diocles starts, like Archimedes, from the property, proved in Prop. 2, that, if the plane of section cut a diameter AA' of the sphere at right angles in M, and if H, H' be taken on OA, OA' produced respectively so that

$$OA' + A'M : A'M = HM : MA,$$
$$OA + AM : AM = H'M : MA',$$

then the cones HBB', $H'BB'$ are respectively equal to the segments ABB', $A'BB'$.

Then, drawing the inference that
$$HA : AM = OA' : A'M,$$
$$H'A' : A'M = OA : AM,$$

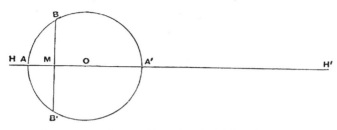

he proceeds to state the problem in the following form, slightly generalising it by the substitution of *any* given straight line for OA or OA':

Given a straight line AA', its extremities A, A', a ratio $C : D$, and another straight line as AK, to divide AA' at M and to find two points H, H' on $A'A$ and AA' produced respectively so that the following relations may hold simultaneously,

$$C : D = HM : MH' \quad \ldots\ldots\ldots(\alpha),$$
$$HA : AM = AK : A'M \quad \ldots\ldots\ldots(\beta),$$
$$H'A' : A'M = AK : AM \quad \ldots\ldots\ldots(\gamma).$$

Analysis.

Suppose the problem solved and the points M, H, H' all found.

Place AK at right angles to AA', and draw $A'K'$ parallel and equal to AK. Join $KM, K'M$, and produce them to meet $K'A', KA$ respectively in E, F. Join KK', draw EG through E parallel to $A'A$ meeting KF in G, and through M draw QMN parallel to AK meeting EG in Q and KK' in N.

Now $HA : AM = A'K' : A'M$, by (β),

$\qquad\qquad\quad = FA : AM$, by similar triangles,

whence $\qquad HA = FA.$

Similarly $\qquad H'A' = A'E.$

Next,

$FA + AM : A'K' + A'M = AM : A'M$

$\qquad\qquad = AK + AM : EA' + A'M$, by similar triangles.

Therefore
$$(FA + AM).(EA' + A'M) = (KA + AM).(K'A' + A'M).$$
Take AR along AH and $A'R'$ along $A'H'$ such that
$$AR = A'R' = AK.$$
Then, since $FA + AM = HM$, $EA' + A'M = MH'$, we have
$$HM.MH' = RM.MR' \quad\ldots\ldots\ldots\ldots\ldots(\delta).$$
(Thus, if R falls between A and H, R' falls on the side of H' remote from A', and *vice versa*.)

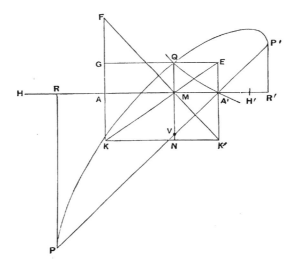

Now $\quad C:D = HM:MH'$, by hypothesis,
$$= HM.MH':MH'^2$$
$$= RM.MR':MH'^2, \text{ by } (\delta).$$
Measure MV along MN so that $MV = A'M$. Join $A'V$ and produce it both ways. Draw RP, $R'P'$ perpendicular to RR' meeting $A'V$ produced in P, P' respectively. Then, the angle $MA'V$ being half a right angle, PP' is given in position, and, since R, R' are given, so are P, P'.

And, by parallels,
$$P'V:PV = R'M:MR.$$

Therefore $PV \cdot P'V : PV^2 = RM \cdot MR' : RM^2$.
But $PV^2 = 2RM^2$.
Therefore $PV \cdot P'V = 2RM \cdot MR'$.
And it was shown that
$$RM \cdot MR' : MH'^2 = C : D.$$
Hence $PV \cdot P'V : MH'^2 = 2C : D$.
But $MH' = A'M + A'E = VM + MQ = QV$.
Therefore $QV^2 : PV \cdot P'V = D : 2C$, a given ratio.

Thus, if we take a line p such that
$$D : 2C = p : PP'*,$$
and if we describe an ellipse with PP' as a diameter and p as the corresponding parameter $[= DD'^2/PP'$ in the ordinary notation of geometrical conics], and such that the ordinates to PP' are inclined to it at an angle equal to half a right angle, i.e. are parallel to QV or AK, then the ellipse will pass through Q.

Hence Q lies on an ellipse given in position.

Again, since EK is a diagonal of the parallelogram GK',
$$GQ \cdot QN = AA' \cdot A'K'.$$

If therefore a rectangular hyperbola be described with KG, KK' as asymptotes and passing through A', it will also pass through Q.

Hence Q lies on a given rectangular hyperbola.

Thus Q is determined as the intersection of a given ellipse

---

* There is a mistake in the Greek text here which seems to have escaped the notice of all the editors up to the present. The words are ἐὰν ἄρα ποιήσωμεν, ὡς τὴν Δ πρὸς τὴν διπλασίαν τῆς Γ, οὕτως τὴν ΤΥ πρὸς ἄλλην τινὰ ὡς τὴν Φ, i.e. (with the lettering above) "If we take a length p such that $D : 2C = PP' : p$." This cannot be right, because we should then have
$$QV^2 : PV \cdot P'V = PP' : p,$$
whereas the two latter terms should be reversed, the correct property of the ellipse being
$$QV^2 : PV \cdot P'V = p : PP'. \qquad \text{[Apollonius I. 21]}$$
The mistake would appear to have originated as far back as Eutocius, but I think that Eutocius is more likely to have made the slip than Diocles himself, because any intelligent mathematician would be more likely to make such a slip in writing out another man's work than to overlook it if made by another.

and a given hyperbola, and is therefore given. Thus M is given, and H, H' can at once be found.

Synthesis.

Place AA', AK at right angles, draw $A'K'$ parallel and equal to AK, and join KK'.

Make AR (measured along $A'A$ produced) and $A'R'$ (measured along AA' produced) each equal to AK, and through R, R' draw perpendiculars to RR'.

Then through A' draw PP' making an angle $(AA'P)$ with AA' equal to half a right angle and meeting the perpendiculars just drawn in P, P' respectively.

Take a length p such that
$$D : 2C = p : PP'*,$$
and with PP' as diameter and p as the corresponding parameter describe an ellipse such that the ordinates to PP' are inclined to it at an angle equal to $AA'P$, i.e. are parallel to AK.

With asymptotes KA, KK' draw a rectangular hyperbola passing through A'.

Let the hyperbola and ellipse meet in Q, and from Q draw $QMVN$ perpendicular to AA' meeting AA' in M, PP' in V and KK' in N. Also draw GQE parallel to AA' meeting AK, $A'K'$ respectively in G, E.

Produce KA, $K'M$ to meet in F.

Then, from the property of the hyperbola,
$$GQ \cdot QN = AA' \cdot A'K',$$
and, since these rectangles are equal, KME is a straight line.

Measure AH along AR equal to AF, and $A'H'$ along $A'R'$ equal to $A'E$.

From the property of the ellipse,
$$QV^2 : PV \cdot P'V = p : PP'$$
$$= D : 2C.$$

---

* Here too the Greek text repeats the same error as that noted on p. 77.

And, by parallels,
$$PV : P'V = RM : R'M,$$
or $\quad PV . P'V : P'V^2 = RM . MR' : R'M^2,$

while $P'V^2 = 2R'M^2$, since the angle $RA'P$ is half a right angle.

Therefore $\quad PV . P'V = 2RM . MR',$

whence $\quad QV^2 : 2RM . MR' = D : 2C.$

But $\quad QV = EA' + A'M = MH'.$

Therefore $\quad RM . MR' : MH'^2 = C : D.$

Again, by similar triangles,
$$FA + AM : K'A' + A'M = AM : A'M$$
$$= KA + AM : EA' + A'M.$$

Therefore
$$(FA + AM).(EA' + A'M) = (KA + AM).(K'A' + A'M)$$
or $\quad HM . MH' = RM . MR'.$

It follows that
$$HM . MH' : MH'^2 = C : D,$$
or $\quad HM : MH' = C : D$ .... (α).

Also $\quad HA : AM = FA : AM,$
$$= A'K' : A'M, \text{ by similar triangles}\ldots(\beta),$$

and $\quad H'A' : A'M = EA' : A'M$
$$= AK : AM \ldots\ldots\ldots\ldots\ldots\ldots (\gamma).$$

Hence the points M, H, H' satisfy the three given relations.]

Proposition 5. (Problem.)

To construct a segment of a sphere similar to one segment and equal in volume to another.

Let ABB' be one segment whose vertex is A and whose base is the circle on BB' as diameter; and let DEF be another segment whose vertex is D and whose base is the circle on EF

80   ARCHIMEDES

as diameter. Let AA', DD' be diameters of the great circles passing through BB', EF respectively, and let O, C be the respective centres of the spheres.

Suppose it required to draw a segment similar to DEF and equal in volume to ABB'.

Analysis. Suppose the problem solved, and let def be the required segment, d being the vertex and ef the diameter of the base. Let dd' be the diameter of the sphere which bisects ef at right angles, c the centre of the sphere.

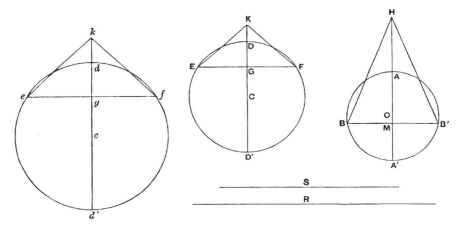

Let M, G, g be the points where BB', EF, ef are bisected at right angles by AA', DD', dd' respectively, and produce OA, CD, cd respectively to H, K, k, so that

$$\left.\begin{array}{l} OA' + A'M : A'M = HM : MA \\ CD' + D'G : D'G = KG : GD \\ cd' + d'g : d'g = kg : gd \end{array}\right\},$$

and suppose cones formed with vertices H, K, k and with the same bases as the respective segments. The cones will then be equal to the segments respectively [Prop. 2].

Therefore, by hypothesis,

the cone HBB' = the cone kef.

ON THE SPHERE AND CYLINDER II.

Hence

(circle on diameter BB') : (circle on diameter ef) = $kg : HM$,

so that $\qquad BB'^2 : ef^2 = kg : HM$ .... (1).

But, since the segments DEF, def are similar, so are the cones KEF, kef.

Therefore $\qquad KG : EF = kg : ef$.

And the ratio $KG : EF$ is given. Therefore the ratio $kg : ef$ is given.

Suppose a length R taken such that

$$kg : ef = HM : R \quad \ldots\ldots\ldots\ldots\ldots\ldots (2).$$

Thus R is given.

Again, since $kg : HM = BB'^2 : ef^2 = ef : R$, by (1) and (2), suppose a length S taken such that

$$ef^2 = BB' . S,$$

or $\qquad BB'^2 : ef^2 = BB' : S$.

Thus $\qquad BB' : ef = ef : S = S : R$,

and ef, S *are two mean proportionals in continued proportion between* BB', R.

Synthesis. Let ABB', DEF be great circles, AA', DD' the diameters bisecting BB', EF at right angles in M, G respectively, and O, C the centres.

Take H, K in the same way as before, and construct the cones HBB', KEF, which are therefore equal to the respective segments ABB', DEF.

Let R be a straight line such that

$$KG : EF = HM : R,$$

and between BB', R take two mean proportionals ef, S.

On ef as base describe a segment of a circle with vertex d and similar to the segment of a circle DEF. Complete the circle, and let dd' be the diameter through d, and c the centre. Conceive a sphere constructed of which def is a great circle, and through ef draw a plane at right angles to dd'.

Then shall *def* be the required segment of a sphere.

For the segments DEF, def of the spheres are similar, like the circular segments DEF, def.

Produce cd to k so that

$$cd' + d'g : d'g = kg : gd.$$

The cones KEF, kef are then similar.

Therefore $kg : ef = KG : EF = HM : R$,

whence $kg : HM = ef : R$.

But, since BB', ef, S, R are in continued proportion,

$$BB'^2 : ef^2 = BB' : S$$
$$= ef : R$$
$$= kg : HM.$$

Thus the bases of the cones HBB', kef are reciprocally proportional to their heights. The cones are therefore equal, and *def* is the segment required, being equal in volume to the cone *kef*. [Prop. 2]

Proposition 6. (Problem.)

Given two segments of spheres, to find a third segment of a sphere similar to one of the given segments and having its surface equal to that of the other.

Let ABB' be the segment to whose surface the surface of the required segment is to be equal, $ABA'B'$ the great circle whose plane cuts the plane of the base of the segment ABB' at right angles in BB'. Let AA' be the diameter which bisects BB' at right angles.

Let DEF be the segment to which the required segment is to be similar, $DED'F$ the great circle cutting the base of the segment at right angles in EF. Let DD' be the diameter bisecting EF at right angles in G.

Suppose the problem solved, *def* being a segment similar to DEF and having its surface equal to that of ABB'; and complete the figure for *def* as for *DEF*, corresponding points being denoted by small and capital letters respectively.

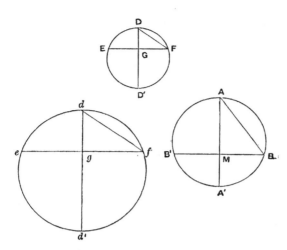

Join AB, DF, df.

Now, since the surfaces of the segments def, ABB' are equal, so are the circles on df, AB as diameters;   [I. 42, 43]

that is, $df = AB$.

From the similarity of the segments DEF, def we obtain
$$d'd : dg = D'D : DG,$$
and $$dg : df = DG : DF;$$
whence $$d'd : df = D'D : DF,$$
or $$d'd : AB = D'D : DF.$$

But AB, $D'D$, DF are all given;

therefore $d'd$ is given.

Accordingly the synthesis is as follows.

Take $d'd$ such that
$$d'd : AB = D'D : DF \quad\quad\quad\quad\ldots\ldots\ldots\ldots\ldots\ldots(1).$$

Describe a circle on $d'd$ as diameter, and conceive a sphere constructed of which this circle is a great circle.

Divide $d'd$ at g so that
$$d'g : gd = D'G : GD,$$
and draw through g a plane perpendicular to $d'd$ cutting off the segment def of the sphere and intersecting the plane of the great circle in ef. The segments def, DEF are thus similar, and
$$dg : df = DG : DF.$$

But from above, *componendo*,
$$d'd : dg = D'D : DG.$$
Therefore, *ex aequali*, $d'd : df = D'D : DF,$
whence, by (1), $df = AB$.

Therefore the segment def has its surface equal to the surface of the segment ABB' [I. 42, 43], while it is also similar to the segment DEF.

Proposition 7. (Problem.)

From a given sphere to cut off a segment by a plane so that the segment may have a given ratio to the cone which has the same base as the segment and equal height.

Let AA' be the diameter of a great circle of the sphere. It is required to draw a plane at right angles to AA' cutting off a segment, as ABB', such that the segment ABB' has to the cone ABB' a given ratio.

Analysis.

Suppose the problem solved, and let the plane of section cut the plane of the great circle in BB', and the diameter AA' in M. Let O be the centre of the sphere.

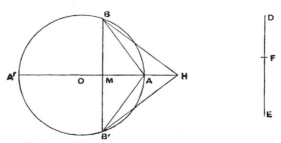

Produce OA to H so that
$$OA' + A'M : A'M = HM : MA \quad \ldots\ldots\ldots\ldots(1).$$

Thus the cone HBB' is equal to the segment ABB'. [Prop. 2]

Therefore the given ratio must be equal to the ratio of the cone HBB' to the cone ABB', *i.e.* to the ratio $HM : MA$.

Hence the ratio $OA' + A'M : A'M$ is given; and therefore $A'M$ is given.

διορισμός.

Now $OA' : A'M > OA' : A'A$,

so that $OA' + A'M : A'M > OA' + A'A : A'A$
$> 3 : 2$.

Thus, *in order that a solution may be possible, it is a necessary condition that the given ratio must be greater than* 3 : 2.

The **synthesis** proceeds thus.

Let AA' be a diameter of a great circle of the sphere, O the centre.

Take a line DE, and a point F on it, such that $DE : EF$ is equal to the given ratio, being greater than 3 : 2.

Now, since $OA' + A'A : A'A = 3 : 2$,

$DE : EF > OA' + A'A : A'A$,

so that $DF : FE > OA' : A'A$.

Hence a point M can be found on AA' such that

$DF : FE = OA' : A'M$ ....(2).

Through M draw a plane at right angles to AA' intersecting the plane of the great circle in BB', and cutting off from the sphere the segment ABB'.

As before, take H on OA produced such that

$OA' + A'M : A'M = HM : MA$.

Therefore $HM : MA = DE : EF$, by means of (2).

It follows that the cone HBB', or the segment ABB', is to the cone ABB' in the given ratio $DE : EF$.

Proposition 8.

If a sphere be cut by a plane not passing through the centre into two segments $A'BB'$, ABB', of which $A'BB'$ is the greater, then the ratio

$(segmt.\ A'BB') : (segmt.\ ABB')$
$< (surface\ of\ A'BB')^2 : (surface\ of\ ABB')^2$
but $> (surface\ of\ A'BB')^{\frac{3}{2}} : (surface\ of\ ABB')^{\frac{3}{2}}$*.

Let the plane of section cut a great circle $A'BAB'$ at right angles in BB', and let AA' be the diameter bisecting BB' at right angles in M.

Let O be the centre of the sphere.

Join $A'B$, AB.

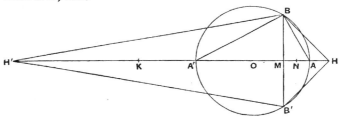

As usual, take H on OA produced, and H' on OA' produced, so that

$$OA' + A'M : A'M = HM : MA \quad\quad\quad (1),$$
$$OA + AM : AM = H'M : MA' \quad\quad\quad (2),$$

and conceive cones drawn each with the same base as the two segments and with apices H, H' respectively. The cones are then respectively equal to the segments [Prop. 2], and they are in the ratio of their heights HM, $H'M$.

Also

(surface of $A'BB'$) : (surface of ABB') $= A'B^2 : AB^2$ [I. 42, 43]
$\quad\quad\quad\quad\quad\quad\quad\quad\quad\quad\quad\quad\quad = A'M : AM.$

---

* This is expressed in Archimedes' phrase by saying that the greater segment has to the lesser a ratio "less than the duplicate ($\delta\iota\pi\lambda\acute{a}\sigma\iota o\nu$) of that which the surface of the greater segment has to the surface of the lesser, but greater than the sesquialterate ($\dot{\eta}\mu\iota\acute{o}\lambda\iota o\nu$) [of that ratio]."

We have therefore to prove

(a) that $\quad H'M : MH < A'M^2 : MA^2$,

(b) that $\quad H'M : MH > A'M^2 : MA^2$.

(a) From (2) above,

$$A'M : AM = H'M : OA + AM$$
$$= H'A' : OA', \text{ since } OA = OA'.$$

Since $A'M > AM$, $H'A' > OA'$; therefore, if we take K on $H'A'$ so that $OA' = A'K$, K will fall between H' and A'.

And, by (1), $\quad A'M : AM = KM : MH$.

Thus $\quad KM : MH = H'A' : A'K$, since $A'K = OA'$,
$$> H'M : MK.$$

Therefore $\quad H'M \cdot MH < KM^2$.

It follows that

$$H'M \cdot MH : MH^2 < KM^2 : MH^2,$$

or $\quad H'M : MH < KM^2 : MH^2$
$$< A'M^2 : AM^2, \text{ by (1)}.$$

(b) Since $OA' = OA$,

$$A'M \cdot MA < A'O \cdot OA,$$

or $\quad A'M : OA' < OA : AM$
$$< H'A' : A'M, \text{ by means of (2)}.$$

Therefore $\quad A'M^2 < H'A' \cdot OA'$
$$< H'A' \cdot A'K.$$

Take a point N on $A'A$ such that

$$A'N^2 = H'A' \cdot A'K.$$

Thus $\quad H'A' : A'K = A'N^2 : A'K^2 \quad \ldots\ldots\ldots\ldots\ldots(3)$.

Also $\quad H'A' : A'N = A'N : A'K$,

and, componendo,

$$H'N : A'N = NK : A'K,$$

whence $\quad A'N^2 : A'K^2 = H'N^2 : NK^2$.

Therefore, by (3),

$$H'A' : A'K = H'N^2 : NK^2.$$

Now $H'M : MK > H'N : NK.$

Therefore
$$H'M^2 : MK^2 > H'A' : A'K$$
$$> H'A' : OA'$$
$$> A'M : MA, \text{ by (2), as above,}$$
$$> OA' + A'M : MH, \text{ by (1),}$$
$$> KM : MH.$$

Hence $H'M^2 : MH^2 = (H'M^2 : MK^2) \cdot (KM^2 : MH^2)$
$$> (KM : MH) \cdot (KM^2 : MH^2).$$

It follows that

$$H'M : MH > KM^{\frac{3}{2}} : MH^{\frac{3}{2}}$$
$$> A'M^{\frac{3}{2}} : AM^{\frac{3}{2}}, \text{ by (1)}.$$

[The text of Archimedes adds an alternative proof of this proposition, which is here omitted because it is in fact neither clearer nor shorter than the above.]

Proposition 9.

Of all segments of spheres which have equal surfaces the hemisphere is the greatest in volume.

Let $ABA'B'$ be a great circle of a sphere, AA' being a diameter, and O the centre. Let the sphere be cut by a plane, not passing through O, perpendicular to AA' (at M), and intersecting the plane of the great circle in BB'. The segment ABB' may then be either less than a hemisphere as in Fig. 1, or greater than a hemisphere as in Fig. 2.

Let $DED'E'$ be a great circle of another sphere, DD' being a diameter and C the centre. Let the sphere be cut by a plane through C perpendicular to DD' and intersecting the plane of the great circle in the diameter EE'.

Suppose the surfaces of the segment ABB' and of the hemisphere DEE' to be equal.

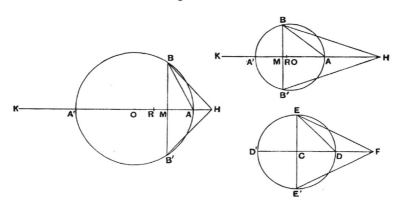

Since the surfaces are equal, $AB = DE$. [I. 42, 43]

Now, in Fig. 1, $AB^2 > 2AM^2$ and $< 2AO^2$,

and, in Fig. 2, $AB^2 < 2AM^2$ and $> 2AO^2$.

Hence, if R be taken on AA' such that
$$AR^2 = \tfrac{1}{2}AB^2,$$
R will fall between O and M.

Also, since $AB^2 = DE^2$, $AR = CD$.

Produce OA' to K so that $OA' = A'K$, and produce $A'A$ to H so that
$$A'K : A'M = HA : AM,$$
or, *componendo*, $A'K + A'M : A'M = HM : MA$ ....(1).

Thus the cone HBB' is equal to the segment ABB'.
[Prop. 2]

Again, produce CD to F so that $CD = DF$, and the cone FEE' will be equal to the hemisphere DEE'. [Prop. 2]

Now $AR \cdot RA' > AM \cdot MA'$,

and $AR^2 = \tfrac{1}{2}AB^2 = \tfrac{1}{2}AM \cdot AA' = AM \cdot A'K$.

Hence
$$AR \cdot RA' + RA^2 > AM \cdot MA' + AM \cdot A'K,$$
or
$$AA' \cdot AR > AM \cdot MK$$
$$> HM \cdot A'M, \text{ by (1)}.$$

Therefore $AA' : A'M > HM : AR,$

or $AB^2 : BM^2 > HM : AR,$

i.e. $AR^2 : BM^2 > HM : 2AR$, since $AB^2 = 2AR^2$,
$$> HM : CF.$$

Thus, since $AR = CD$, or CE,

(circle on diam. EE') : (circle on diam. BB') > $HM : CF$.

It follows that

(the cone FEE') > (the cone HBB'),

and therefore the hemisphere DEE' is greater in volume than the segment ABB'.

MEASUREMENT OF A CIRCLE.

Proposition 1.

The area of any circle is equal to a right-angled triangle in which one of the sides about the right angle is equal to the radius, and the other to the circumference, of the circle.

Let $ABCD$ be the given circle, K the triangle described.

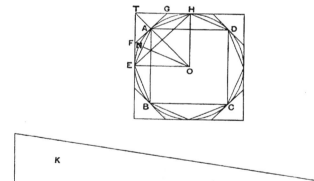

Then, if the circle is not equal to K, it must be either greater or less.

I. If possible, let the circle be greater than K.

Inscribe a square $ABCD$, bisect the arcs AB, BC, CD, DA, then bisect (if necessary) the halves, and so on, until the sides of the inscribed polygon whose angular points are the points of division subtend segments whose sum is less than the excess of the area of the circle over K.

Thus the area of the polygon is greater than K.

Let AE be any side of it, and ON the perpendicular on AE from the centre O.

Then ON is less than the radius of the circle and therefore less than one of the sides about the right angle in K. Also the perimeter of the polygon is less than the circumference of the circle, i.e. less than the other side about the right angle in K.

Therefore the area of the polygon is less than K; which is inconsistent with the hypothesis.

Thus the area of the circle is not greater than K.

II. If possible, let the circle be less than K.

Circumscribe a square, and let two adjacent sides, touching the circle in E, H, meet in T. Bisect the arcs between adjacent points of contact and draw the tangents at the points of bisection. Let A be the middle point of the arc EH, and FAG the tangent at A.

Then the angle TAG is a right angle.

Therefore $\qquad TG > GA$
$\qquad\qquad\quad > GH.$

It follows that the triangle FTG is greater than half the area $TEAH$.

Similarly, if the arc AH be bisected and the tangent at the point of bisection be drawn, it will cut off from the area GAH more than one-half.

Thus, by continuing the process, we shall ultimately arrive at a circumscribed polygon such that the spaces intercepted between it and the circle are together less than the excess of K over the area of the circle.

Thus the area of the polygon will be less than K.

Now, since the perpendicular from O on any side of the polygon is equal to the radius of the circle, while the perimeter of the polygon is greater than the circumference of the circle, it follows that the area of the polygon is greater than the triangle K; which is impossible.

Therefore the area of the circle is not less than K.

Since then the area of the circle is neither greater nor less than K, it is equal to it.

Proposition 2.

The area of a circle is to the square on its diameter as 11 to 14.

[The text of this proposition is not satisfactory, and Archimedes cannot have placed it before Proposition 3, as the approximation depends upon the result of that proposition.]

Proposition 3.

The ratio of the circumference of any circle to its diameter is less than $3\frac{1}{7}$ but greater than $3\frac{10}{71}$.

[In view of the interesting questions arising out of the arithmetical content of this proposition of Archimedes, it is necessary, in reproducing it, to distinguish carefully the actual steps set out in the text as we have it from the intermediate steps (mostly supplied by Eutocius) which it is convenient to put in for the purpose of making the proof easier to follow. Accordingly all the steps not actually appearing in the text have been enclosed in square brackets, in order that it may be clearly seen how far Archimedes omits actual calculations and only gives results. It will be observed that he gives two fractional approximations to $\sqrt{3}$ (one being less and the other greater than the real value) without any explanation as to how he arrived at them; and in like manner approximations to the square roots of several large numbers which are not complete squares are merely stated. These various approximations and the machinery of Greek arithmetic in general will be found discussed in the Introduction, Chapter IV.]

I. Let AB be the diameter of any circle, O its centre, AC the tangent at A; and let the angle AOC be one-third of a right angle.

Then $OA : AC\, [=\sqrt{3} : 1] > 265 : 153$ .... (1),

and $OC : CA\, [=2 : 1] = 306 : 153$ .... (2).

First, draw OD bisecting the angle AOC and meeting AC in D.

Now $CO : OA = CD : DA$, [Eucl. VI. 3]

so that $[CO + OA : OA = CA : DA$, or]

$CO + OA : CA = OA : AD.$

Therefore [by (1) and (2)]

$OA : AD > 571 : 153$ .... (3).

Hence $OD^2 : AD^2\, [= (OA^2 + AD^2) : AD^2$

$> (571^2 + 153^2) : 153^2]$

$> 349450 : 23409,$

so that $OD : DA > 591\tfrac{1}{8} : 153$ .... (4).

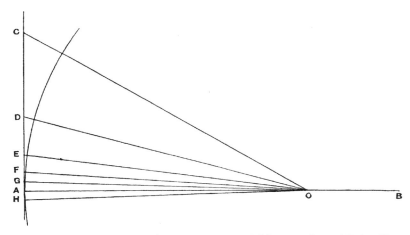

Secondly, let OE bisect the angle AOD, meeting AD in E.

[Then $DO : OA = DE : EA,$

so that $DO + OA : DA = OA : AE.$]

Therefore $OA : AE\, [> (591\tfrac{1}{8} + 571) : 153$, by (3) and (4)]

$> 1162\tfrac{1}{8} : 153$ .... (5).

[It follows that
$$OE^2 : EA^2 > \{(1162\tfrac{1}{8})^2 + 153^2\} : 153^2$$
$$> (1350534\tfrac{33}{64} + 23409) : 23409$$
$$> 1373943\tfrac{33}{64} : 23409.]$$

Thus $\qquad OE : EA > 1172\tfrac{1}{8} : 153$ ....(6).

Thirdly, let OF bisect the angle AOE and meet AE in F.

We thus obtain the result [corresponding to (3) and (5) above] that
$$OA : AF [> (1162\tfrac{1}{8} + 1172\tfrac{1}{8}) : 153]$$
$$> 2334\tfrac{1}{4} : 153 ....(7).$$

[Therefore $\quad OF^2 : FA^2 > \{(2334\tfrac{1}{4})^2 + 153^2\} : 153^2$
$$> 5472132\tfrac{1}{16} : 23409.]$$

Thus $\qquad OF : FA > 2339\tfrac{1}{4} : 153$ ....(8).

Fourthly, let OG bisect the angle AOF, meeting AF in G.

We have then

$OA : AG [> (2334\tfrac{1}{4} + 2339\tfrac{1}{4}) : 153$, by means of (7) and (8)]
$$> 4673\tfrac{1}{2} : 153.$$

Now the angle AOC, which is one-third of a right angle, has been bisected four times, and it follows that

$$\angle AOG = \tfrac{1}{48} \text{ (a right angle)}.$$

Make the angle AOH on the other side of OA equal to the angle AOG, and let GA produced meet OH in H.

Then $\qquad \angle GOH = \tfrac{1}{24}$ (a right angle).

Thus GH is one side of a regular polygon of 96 sides circumscribed to the given circle.

And, since $\quad OA : AG > 4673\tfrac{1}{2} : 153,$
while $\qquad AB = 2OA, \quad GH = 2AG,$
it follows that

AB : (perimeter of polygon of 96 sides) $[> 4673\tfrac{1}{2} : 153 \times 96]$
$$> 4673\tfrac{1}{2} : 14688.$$

But
$$\frac{14688}{4673\tfrac{1}{2}} = 3 + \frac{667\tfrac{1}{2}}{4673\tfrac{1}{2}}$$
$$\left[< 3 + \frac{667\tfrac{1}{2}}{4672\tfrac{1}{2}} \right]$$
$$< 3\tfrac{1}{7}.$$

Therefore the circumference of the circle (being less than the perimeter of the polygon) is *a fortiori* less than $3\tfrac{1}{7}$ times the diameter AB.

II. Next let AB be the diameter of a circle, and let AC, meeting the circle in C, make the angle CAB equal to one-third of a right angle. Join BC.

Then $AC : CB \,[= \sqrt{3} : 1] < 1351 : 780.$

First, let AD bisect the angle BAC and meet BC in d and the circle in D. Join BD.

Then $\angle BAD = \angle dAC$
$= \angle dBD,$

and the angles at D, C are both right angles.

It follows that the triangles ADB, $[ACd]$, BDd are similar.

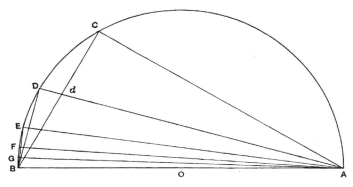

Therefore $AD : DB = BD : Dd$
$[= AC : Cd]$
$= AB : Bd$ [Eucl. VI. 3]
$= AB + AC : Bd + Cd$
$= AB + AC : BC$

or $BA + AC : BC = AD : DB.$

MEASUREMENT OF A CIRCLE.

[But $AC : CB < 1351 : 780$, from above,
while $BA : BC = 2 : 1$
$= 1560 : 780$.]

Therefore $AD : DB < 2911 : 780$ ....(1).

[Hence $AB^2 : BD^2 < (2911^2 + 780^2) : 780^2$
$< 9082321 : 608400$.]

Thus $AB : BD < 3013\frac{3}{4} : 780$ .... (2).

Secondly, let AE bisect the angle BAD, meeting the circle in E; and let BE be joined.

Then we prove, in the same way as before, that
$AE : EB \,[= BA + AD : BD$
$< (3013\frac{3}{4} + 2911) : 780$, by (1) and (2)]
$< 5924\frac{3}{4} : 780$
$< 5924\frac{3}{4} \times \frac{4}{13} : 780 \times \frac{4}{13}$
$< 1823 : 240$ .... (3).

[Hence $AB^2 : BE^2 < (1823^2 + 240^2) : 240^2$
$< 3380929 : 57600$.]

Therefore $AB : BE < 1838\frac{9}{11} : 240$ ....(4).

Thirdly, let AF bisect the angle BAE, meeting the circle in F.

Thus $AF : FB \,[= BA + AE : BE$
$< 3661\frac{9}{11} : 240$, by (3) and (4)]
$< 3661\frac{9}{11} \times \frac{11}{40} : 240 \times \frac{11}{40}$
$< 1007 : 66$ ....(5).

[It follows that
$AB^2 : BF^2 < (1007^2 + 66^2) : 66^2$
$< 1018405 : 4356$.]

Therefore $AB : BF < 1009\frac{1}{6} : 66$ ....(6).

Fourthly, let the angle BAF be bisected by AG meeting the circle in G.

Then $AG : GB \,[= BA + AF : BF]$
$< 2016\frac{1}{6} : 66$, by (5) and (6).

[And $AB^2 : BG^2 < \{(2016\tfrac{1}{6})^2 + 66^2\} : 66^2$
$< 4069284\tfrac{1}{36} : 4356.$]

Therefore $AB : BG < 2017\tfrac{1}{4} : 66,$

whence $BG : AB > 66 : 2017\tfrac{1}{4}$ ....(7).

[Now the angle BAG which is the result of the fourth bisection of the angle BAC, or of one-third of a right angle, is equal to one-fortyeighth of a right angle.

Thus the angle subtended by BG at the centre is

$\tfrac{1}{24}$ (a right angle).]

Therefore BG is a side of a regular inscribed polygon of 96 sides.

It follows from (7) that

(perimeter of polygon) : AB [$> 96 \times 66 : 2017\tfrac{1}{4}$]

$> 6336 : 2017\tfrac{1}{4}.$

And $\dfrac{6336}{2017\tfrac{1}{4}} > 3\tfrac{10}{71}.$

Much more then is the circumference of the circle greater than $3\tfrac{10}{71}$ times the diameter.

Thus the ratio of the circumference to the diameter

$< 3\tfrac{1}{7}$ but $> 3\tfrac{10}{71}.$

ON CONOIDS AND SPHEROIDS.

Introduction*.

"ARCHIMEDES to Dositheus greeting.

In this book I have set forth and send you the proofs of the remaining theorems not included in what I sent you before, and also of some others discovered later which, though I had often tried to investigate them previously, I had failed to arrive at because I found their discovery attended with some difficulty. And this is why even the propositions themselves were not published with the rest. But afterwards, when I had studied them with greater care, I discovered what I had failed in before.

Now the remainder of the earlier theorems were propositions concerning the right-angled conoid [paraboloid of revolution]; but the discoveries which I have now added relate to an obtuse-angled conoid [hyperboloid of revolution] and to spheroidal figures, some of which I call *oblong* (παραμάκεα) and others *flat* (ἐπιπλατέα).

I. Concerning the *right-angled conoid* it was laid down that, if a section of a right-angled cone [a parabola] be made to revolve about the diameter [axis] which remains fixed and

---

* The whole of this introductory matter, including the definitions, is translated literally from the Greek text in order that the terminology of Archimedes may be faithfully represented. When this has once been set out, nothing will be lost by returning to modern phraseology and notation. These will accordingly be employed, as usual, when we come to the actual propositions of the treatise.

return to the position from which it started, the figure comprehended by the section of the right-angled cone is called a **right-angled conoid,** and the diameter which has remained fixed is called its **axis,** while its **vertex** is the point in which the axis meets ($ἅπτεται$) the surface of the conoid. And if a plane touch the right-angled conoid, and another plane drawn parallel to the tangent plane cut off a segment of the conoid, the **base** of the segment cut off is defined as the portion intercepted by the section of the conoid on the cutting plane, the **vertex** [of the segment] as the point in which the first plane touches the conoid, and the **axis** [of the segment] as the portion cut off within the segment from the line drawn through the vertex of the segment parallel to the axis of the conoid.

The questions propounded for consideration were

(1) why, if a segment of the right-angled conoid be cut off by a plane at right angles to the axis, will the segment so cut off be half as large again as the cone which has the same base as the segment and the same axis, and

(2) why, if two segments be cut off from the right-angled conoid by planes drawn in any manner, will the segments so cut off have to one another the duplicate ratio of their axes.

II. Respecting the *obtuse-angled conoid* we lay down the following premisses. If there be in a plane a section of an obtuse-angled cone [a hyperbola], its diameter [axis], and the nearest lines to the section of the obtuse-angled cone [*i.e.* the asymptotes of the hyperbola], and if, the diameter [axis] remaining fixed, the plane containing the aforesaid lines be made to revolve about it and return to the position from which it started, the nearest lines to the section of the obtuse-angled cone [the asymptotes] will clearly comprehend an isosceles cone whose vertex will be the point of concourse of the nearest lines and whose axis will be the diameter [axis] which has remained fixed. The figure comprehended by the section of the obtuse-angled cone is called an **obtuse-angled conoid** [hyperboloid of revolution], its **axis** is the diameter which has remained fixed, and its **vertex** the point in which the axis meets the surface of the conoid. The cone comprehended by the nearest lines to the section of the obtuse-angled cone is called [the cone] **enveloping the conoid** (περιέχων τὸ κωνοειδές), and the straight line between the vertex of the conoid and the vertex of the cone enveloping the conoid is called [the line] **adjacent to the axis** (ποτεοῦσα τῷ ἄξονι). And if a plane touch the obtuse-angled conoid, and another plane drawn parallel to the tangent plane cut off a segment of the conoid, the **base** of the segment so cut off is defined as the portion intercepted by the section of the conoid on the cutting plane, the **vertex** [of the segment] as the point of contact of the plane which touches the conoid, the **axis** [of the segment] as the portion cut off within the segment from the line drawn through the vertex of the segment and the vertex of the cone enveloping the conoid; and the straight line between the said vertices is called **adjacent to the axis.**

Right-angled conoids are all similar; but of obtuse-angled conoids let those be called similar in which the cones enveloping the conoids are similar.

The following questions are propounded for consideration,

(1) why, if a segment be cut off from the obtuse-angled conoid by a plane at right angles to the axis, the segment so cut off has to the cone which has the same base as the segment and the same axis the ratio which the line equal to the sum of the axis of the segment and three times the line adjacent to the axis bears to the line equal to the sum of the axis of the segment and twice the line adjacent to the axis, and

(2) why, if a segment of the obtuse-angled conoid be cut off by a plane not at right angles to the axis, the segment so cut off will bear to the figure which has the same base as the segment and the same axis, being a segment of a cone* (ἀπότμαμα κώνου), the ratio which the line equal to the sum of the axis of the segment and three times the line adjacent to the axis bears to the line equal to the sum of the axis of the segment and twice the line adjacent to the axis.

---

* A *segment of a cone* is defined later (p. 104).

III. Concerning spheroidal figures we lay down the following premisses. If a section of an acute-angled cone [ellipse] be made to revolve about the greater diameter [major axis] which remains fixed and return to the position from which it started, the figure comprehended by the section of the acute-angled cone is called an **oblong spheroid** ($\pi\alpha\rho\alpha\mu\hat{\alpha}\kappa\epsilon\varsigma$ $\sigma\phi\alpha\iota\rho\omicron\epsilon\iota\delta\acute{\epsilon}\varsigma$). But if the section of the acute-angled cone revolve about the lesser diameter [minor axis] which remains fixed and return to the position from which it started, the figure comprehended by the section of the acute-angled cone is called a **flat spheroid** ($\epsilon\pi\iota\pi\lambda\alpha\tau\grave{\upsilon}$ $\sigma\phi\alpha\iota\rho\omicron\epsilon\iota\delta\acute{\epsilon}\varsigma$). In either of the spheroids the **axis** is defined as the diameter [axis] which has remained fixed, the **vertex** as the point in which the axis meets the surface of the spheroid, the **centre** as the middle point of the axis, and the **diameter** as the line drawn through the centre at right angles to the axis. And, if parallel planes touch, without cutting, either of the spheroidal figures, and if another plane be drawn parallel to the tangent planes and cutting the spheroid, the **base** of the resulting segments is defined as the portion intercepted by the section of the spheroid on the cutting plane, their **vertices** as the points in which the parallel planes touch the spheroid, and their **axes** as the portions cut off within the segments from the straight line joining their vertices. And that the planes touching the spheroid meet its surface at one point only, and that the straight line joining the points of contact passes through the centre of the spheroid, we shall prove. Those spheroidal figures are called **similar** in which the axes have the same ratio to the 'diameters.' And let segments of spheroidal figures and conoids be called **similar** if they are cut off from similar figures and have their bases similar, while their axes, being either at right angles to the planes of the bases or making equal angles with the corresponding diameters [axes] of the bases, have the same ratio to one another as the corresponding diameters [axes] of the bases.

The following questions about spheroids are propounded for consideration,

(1) why, if one of the spheroidal figures be cut by a plane

through the centre at right angles to the axis, each of the resulting segments will be double of the cone having the same base as the segment and the same axis; while, if the plane of section be at right angles to the axis without passing through the centre, (a) the greater of the resulting segments will bear to the cone which has the same base as the segment and the same axis the ratio which the line equal to the sum of half the straight line which is the axis of the spheroid and the axis of the lesser segment bears to the axis of the lesser segment, and (b) the lesser segment bears to the cone which has the same base as the segment and the same axis the ratio which the line equal to the sum of half the straight line which is the axis of the spheroid and the axis of the greater segment bears to the axis of the greater segment;

(2) why, if one of the spheroids be cut by a plane passing through the centre but not at right angles to the axis, each of the resulting segments will be double of the figure having the same base as the segment and the same axis and consisting of a segment of a cone*.

(3) But, if the plane cutting the spheroid be neither through the centre nor at right angles to the axis, (a) the greater of the resulting segments will have to the figure which has the same base as the segment and the same axis the ratio which the line equal to the sum of half the line joining the vertices of the segments and the axis of the lesser segment bears to the axis of the lesser segment, and (b) the lesser segment will have to the figure with the same base as the segment and the same axis the ratio which the line equal to the sum of half the line joining the vertices of the segments and the axis of the greater segment bears to the axis of the greater segment. And the figure referred to is in these cases also a segment of a cone*.

When the aforesaid theorems are proved, there are discovered by means of them many theorems and problems.

Such, for example, are the theorems

(1) that similar spheroids and similar segments both of

---

* See the definition of a *segment of a cone* (ἀπότμαμα κώνου) on p. 104.

spheroidal figures and conoids have to one another the triplicate ratio of their axes, and

(2) that in equal spheroidal figures the squares on the 'diameters' are reciprocally proportional to the axes, and, if in spheroidal figures the squares on the 'diameters' are reciprocally proportional to the axes, the spheroids are equal.

Such also is the problem, From a given spheroidal figure or conoid to cut off a segment by a plane drawn parallel to a given plane so that the segment cut off is equal to a given cone or cylinder or to a given sphere.

After prefixing therefore the theorems and directions (ἐπιτάγματα) which are necessary for the proof of them, I will then proceed to expound the propositions themselves to you. Farewell.

DEFINITIONS.

If a cone be cut by a plane meeting all the sides [generators] of the cone, the section will be either a circle or a section of an acute-angled cone [an ellipse]. If then the section be a circle, it is clear that the segment cut off from the cone towards the same parts as the vertex of the cone will be a cone. But, if the section be a section of an acute-angled cone [an ellipse], let the figure cut off from the cone towards the same parts as the vertex of the cone be called a **segment of a cone.** Let the **base** of the segment be defined as the plane comprehended by the section of the acute-angled cone, its **vertex** as the point which is also the vertex of the cone, and its **axis** as the straight line joining the vertex of the cone to the centre of the section of the acute-angled cone.

And if a cylinder be cut by two parallel planes meeting all the sides [generators] of the cylinder, the sections will be either circles or sections of acute-angled cones [ellipses] equal and similar to one another. If then the sections be circles, it is clear that the figure cut off from the cylinder between the parallel planes will be a cylinder. But, if the sections be sections of acute-angled cones [ellipses], let the figure cut off from the cylinder between the parallel planes be called a **frustum** (τόμος) **of a cylinder**. And let the **bases** of the frustum be defined as the planes comprehended by the sections of the acute-angled cones [ellipses], and the **axis** as the straight line joining the centres of the sections of the acute-angled cones, so that the axis will be in the same straight line with the axis of the cylinder."

Lemma.

If in an ascending arithmetical progression consisting of the magnitudes $A_1, A_2, \ldots A_n$ the common difference be equal to the least term A_1, then

$$n \cdot A_n < 2(A_1 + A_2 + \ldots + A_n),$$
and
$$> 2(A_1 + A_2 + \ldots + A_{n-1}).$$

[The proof of this is given incidentally in the treatise *On Spirals*, Prop. 11. By placing lines side by side to represent the terms of the progression and then producing each so as to make it equal to the greatest term, Archimedes gives the equivalent of the following proof.

If $\quad S_n = A_1 + A_2 + \ldots + A_{n-1} + A_n,$
we have also $\quad S_n = A_n + A_{n-1} + A_{n-2} + \ldots + A_1.$
And $\quad A_1 + A_{n-1} = A_2 + A_{n-2} = \ldots = A_n.$

Therefore $\quad 2S_n = (n+1) A_n,$
whence $\quad n \cdot A_n < 2S_n,$
and $\quad n \cdot A_n > 2S_{n-1}.$

Thus, if the progression is $a, 2a, \ldots na$,
$$S_n = \frac{n(n+1)}{2} a,$$
and $\quad n^2 a < 2S_n,$
but $\quad > 2S_{n-1}.$]

Proposition 1.

If $A_1, B_1, C_1, \ldots K_1$ and $A_2, B_2, C_2, \ldots K_2$ be two series of magnitudes such that

$$\left. \begin{array}{l} A_1 : B_1 = A_2 : B_2, \\ B_1 : C_1 = B_2 : C_2, \textit{ and so on} \end{array} \right\} \ldots\ldots\ldots\ldots (\alpha),$$

and if $A_3, B_3, C_3, \ldots K_3$ and $A_4, B_4, C_4, \ldots K_4$ be two other series such that

$$\left. \begin{array}{l} A_1 : A_3 = A_2 : A_4, \\ B_1 : B_3 = B_2 : B_4, \text{ and so on} \end{array} \right\} \ldots\ldots\ldots\ldots(\beta),$$

then $(A_1 + B_1 + C_1 + \ldots + K_1) : (A_3 + B_3 + C_3 + \ldots + K_3)$
$= (A_2 + B_2 + C_2 + \ldots + K_2) : (A_4 + B_4 + \ldots + K_4).$

The proof is as follows.

Since $\qquad A_3 : A_1 = A_4 : A_2,$

and $\qquad A_1 : B_1 = A_2 : B_2,$

while $\qquad B_1 : B_3 = B_2 : B_4,$

we have, *ex aequali*, $A_3 : B_3 = A_4 : B_4.$
Similarly $\qquad B_3 : C_3 = B_4 : C_4,$ and so on $\left. \right\} \ldots\ldots(\gamma).$

Again, it follows from equations (α) that

$$A_1 : A_2 = B_1 : B_2 = C_1 : C_2 = \ldots.$$

Therefore

$$A_1 : A_2 = (A_1 + B_1 + C_1 + \ldots + K_1) : (A_2 + B_2 + \ldots + K_2),$$

or $(A_1 + B_1 + C_1 + \ldots + K_1) : A_1 = (A_2 + B_2 + C_2 + \ldots + K_2) : A_2;$

and $\qquad A_1 : A_3 = A_2 : A_4,$

while from equations (γ) it follows in like manner that

$$A_3 : (A_3 + B_3 + C_3 + \ldots + K_3) = A_4 : (A_4 + B_4 + C_4 + \ldots + K_4).$$

By the last three equations, *ex aequali*,

$(A_1 + B_1 + C_1 + \ldots + K_1) : (A_3 + B_3 + C_3 + \ldots + K_3)$
$= (A_2 + B_2 + C_2 + \ldots + K_2) : (A_4 + B_4 + C_4 + \ldots + K_4).$

COR. If any terms in the third and fourth series corresponding to terms in the first and second be left out, the result is the same. For example, if the last terms K_3, K_4 are absent,

$(A_1 + B_1 + C_1 + \ldots + K_1) : (A_3 + B_3 + C_3 + \ldots + I_3)$
$= (A_2 + B_2 + C_2 + \ldots + K_2) : (A_4 + B_4 + C_4 + \ldots + I_4),$

where I immediately precedes K in each series.

Lemma to Proposition 2.

[*On Spirals*, Prop. 10.]

If A_1, A_2, A_3, ...A_n be n lines forming an ascending arithmetical progression in which the common difference is equal to the least term A_1, then

$$(n+1)A_n^2 + A_1(A_1 + A_2 + A_3 + ... + A_n)$$
$$= 3(A_1^2 + A_2^2 + A_3^2 + ... + A_n^2).$$

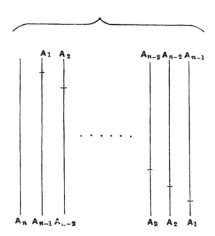

Let the lines A_n, A_{n-1}, A_{n-2}, ...A_1 be placed in a row from left to right. Produce A_{n-1}, A_{n-2}, ...A_1 until they are each equal to A_n, so that the parts produced are respectively equal to A_1, A_2, ...A_{n-1}.

Taking each line successively, we have

$$2A_n^2 = 2A_n^2,$$
$$(A_1 + A_{n-1})^2 = A_1^2 + A_{n-1}^2 + 2A_1 \cdot A_{n-1},$$
$$(A_2 + A_{n-2})^2 = A_2^2 + A_{n-2}^2 + 2A_2 \cdot A_{n-2},$$
$$\dots\dots\dots\dots\dots\dots\dots\dots\dots\dots\dots\dots\dots$$
$$(A_{n-1} + A_1)^2 = A_{n-1}^2 + A_1^2 + 2A_{n-1} \cdot A_1.$$

And, by addition,
$$(n+1)A_n^2 = 2(A_1^2 + A_2^2 + \ldots + A_n^2)$$
$$+ 2A_1 . A_{n-1} + 2A_2 . A_{n-2} + \ldots + 2A_{n-1} . A_1.$$

Therefore, in order to obtain the required result, we have to prove that
$$2(A_1.A_{n-1}+A_2.A_{n-2}+\ldots+A_{n-1}.A_1)+A_1(A_1+A_2+A_3+\ldots+A_n)$$
$$= A_1^2 + A_2^2 + \ldots + A_n^2 \quad \ldots \ldots \ldots \ldots (\alpha).$$

Now $\quad 2A_2 . A_{n-2} = A_1 . 4A_{n-2}$, because $A_2 = 2A_1$,
$\quad\quad\quad 2A_3 . A_{n-3} = A_1 . 6A_{n-3}$, because $A_3 = 3A_1$,
$\quad\quad\quad \ldots\ldots\ldots\ldots\ldots\ldots\ldots\ldots$
$\quad\quad\quad 2A_{n-1} . A_1 = A_1 . 2(n-1)A_1.$

It follows that
$$2(A_1.A_{n-1}+A_2.A_{n-2}+\ldots+A_{n-1}.A_1)+A_1(A_1+A_2+\ldots+A_n)$$
$$= A_1\{A_n + 3A_{n-1} + 5A_{n-2} + \ldots + (2n-1)A_1\}.$$

And this last expression can be proved to be equal to
$$A_1^2 + A_2^2 + \ldots + A_n^2.$$

For $\quad A_n^2 = A_1(n . A_n)$
$\quad\quad\quad\quad = A_1\{A_n + (n-1)A_n\}$
$\quad\quad\quad\quad = A_1\{A_n + 2(A_{n-1} + A_{n-2} + \ldots + A_1)\},$
$\quad\quad\quad$ because $(n-1)A_n = A_{n-1} + A_1$
$\quad\quad\quad\quad\quad\quad\quad\quad\quad + A_{n-2} + A_2$
$\quad\quad\quad\quad\quad\quad\quad\quad\quad + \ldots\ldots\ldots$
$\quad\quad\quad\quad\quad\quad\quad\quad\quad + A_1 + A_{n-1}.$

Similarly $A_{n-1}^2 = A_1\{A_{n-1} + 2(A_{n-2} + A_{n-3} + \ldots + A_1)\},$
$\quad\quad\quad \ldots\ldots\ldots\ldots\ldots\ldots\ldots\ldots$
$\quad\quad\quad A_2^2 = A_1(A_2 + 2A_1),$
$\quad\quad\quad A_1^2 = A_1 . A_1;$

whence, by addition,
$$A_1^2 + A_2^2 + A_3^2 + \ldots + A_n^2$$
$$= A_1\{A_n + 3A_{n-1} + 5A_{n-2} + \ldots + (2n-1)A_1\}.$$

Thus the equation marked (a) above is true; and it follows that

$$(n+1)A_n^2 + A_1(A_1 + A_2 + A_3 + \ldots + A_n) = 3(A_1^2 + A_2^2 + \ldots + A_n^2).$$

COR. 1. *From this it is evident that*

$$n \cdot A_n^2 < 3(A_1^2 + A_2^2 + \ldots + A_n^2) \quad \ldots\ldots\ldots\ldots (1).$$

Also $A_n^2 = A_1\{A_n + 2(A_{n-1} + A_{n-2} + \ldots + A_1)\}$, as above, so that $A_n^2 > A_1(A_n + A_{n-1} + \ldots + A_1)$,

and therefore

$$A_n^2 + A_1(A_1 + A_2 + \ldots + A_n) < 2A_n^2.$$

It follows from the proposition that

$$n \cdot A_n^2 > 3(A_1^2 + A_2^2 + \ldots + A_{n-1}^2) \quad \ldots\ldots\ldots\ldots\ldots (2).$$

COR. 2. All these results will hold if we substitute *similar figures* for squares on all the lines; for similar figures are in the duplicate ratio of their sides.

[In the above proposition the symbols $A_1, A_2, \ldots A_n$ have been used instead of $a, 2a, 3a, \ldots na$ in order to exhibit the geometrical character of the proof; but, if we now substitute the latter terms in the various results, we have (1)

$$(n+1)n^2a^2 + a(a + 2a + \ldots + na)$$
$$= 3\{a^2 + (2a)^2 + (3a)^2 + \ldots + (na)^2\}.$$

Therefore
$$a^2 + (2a)^2 + (3a)^2 + \ldots + (na)^2$$
$$= \frac{a^2}{3}\left\{(n+1)n^2 + \frac{n(n+1)}{2}\right\}$$
$$= a^2 \cdot \frac{n(n+1)(2n+1)}{6}.$$

Also (2) $\quad n^3 < 3(1^2 + 2^2 + 3^2 + \ldots + n^2),$

and (3) $\quad n^3 > 3(1^2 + 2^2 + 3^2 + \ldots + \overline{n-1}|^2).$]

Proposition 2.

If $A_1, A_2 \ldots A_n$ be any number of areas such that[*]
$$A_1 = ax + x^2,$$
$$A_2 = a \cdot 2x + (2x)^2,$$
$$A_3 = a \cdot 3x + (3x)^2,$$
$$\ldots\ldots\ldots\ldots\ldots\ldots$$
$$A_n = a \cdot nx + (nx)^2,$$

then $\quad n \cdot A_n : (A_1 + A_2 + \ldots + A_n) < (a + nx) : \left(\dfrac{a}{2} + \dfrac{nx}{3}\right),$

and $\quad n \cdot A_n : (A_1 + A_2 + \ldots + A_{n-1}) > (a + nx) : \left(\dfrac{a}{2} + \dfrac{nx}{3}\right).$

For, by the Lemma immediately preceding Prop. 1,
$$n \cdot anx < 2(ax + a \cdot 2x + \ldots + a \cdot nx),$$
and $\qquad\qquad > 2(ax + a \cdot 2x + \ldots + a \cdot \overline{n-1}\, x).$

Also, by the Lemma preceding this proposition,
$$n \cdot (nx)^2 < 3\{x^2 + (2x)^2 + (3x)^2 + \ldots + (nx)^2\}$$
and $\qquad\qquad > 3\{x^2 + (2x)^2 + \ldots + (\overline{n-1}\,x)^2\}.$

Hence
$$\dfrac{an^2 x}{2} + \dfrac{n(nx)^2}{3} < [(ax + x^2) + \{a \cdot 2x + (2x)^2\} + \ldots + \{a \cdot nx + (nx)^2\}],$$
and
$$> [(ax + x^2) + \{a \cdot 2x + (2x)^2\} + \ldots + \{a \cdot \overline{n-1}\,x + (\overline{n-1}\,x)^2\}],$$

or $\qquad\qquad \dfrac{an^2 x}{2} + \dfrac{n(nx)^2}{3} < A_1 + A_2 + \ldots + A_n,$

and $\qquad\qquad\qquad\qquad > A_1 + A_2 + \ldots + A_{n-1}.$

It follows that
$$n \cdot A_n : (A_1 + A_2 + \ldots + A_n) < n\{a \cdot nx + (nx)^2\} : \left\{\dfrac{an^2 x}{2} + \dfrac{n(nx)^2}{3}\right\},$$

or $\qquad n \cdot A_n : (A_1 + A_2 + \ldots + A_n) < (a + nx) : \left(\dfrac{a}{2} + \dfrac{nx}{3}\right);$

also $\qquad n \cdot A_n : (A_1 + A_2 + \ldots + A_{n-1}) > (a + nx) : \left(\dfrac{a}{2} + \dfrac{nx}{3}\right).$

---

[*] The phraseology of Archimedes here is that associated with the traditional method of application of areas: εἴ κα…παρ' ἑκάσταν αὐτᾶν παραπέσῃ τι χωρίον ὑπερβάλλον εἴδει τετραγώνῳ, "if to each of the lines there be applied a space [rectangle] exceeding by a square figure." Thus A_1 is a rectangle of height x applied to a line a but overlapping it so that the base extends a distance x beyond a.

Proposition 3.

(1) *If TP, TP' be two tangents to any conic meeting in T, and if Qq, $Q'q'$ be any two chords parallel respectively to TP, TP' and meeting in O, then*

$$QO \cdot Oq : Q'O \cdot Oq' = TP^2 : TP'^2.$$

"And this is proved in the elements of conics*."

(2) *If QQ' be a chord of a parabola bisected in V by the diameter PV, and if PV be of constant length, then the areas of the triangle PQQ' and of the segment PQQ' are both constant whatever be the direction of QQ'.*

Let ABB' be the particular segment of the parabola whose vertex is A, so that BB' is bisected perpendicularly by the axis at the point H, where $AH = PV$.

Draw QD perpendicular to PV.

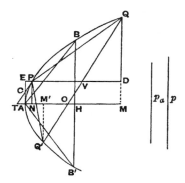

Let p_a be the parameter of the principal ordinates, and let p be another line of such length that

$$QV^2 : QD^2 = p : p_a;$$

it will then follow that p is equal to the parameter of the ordinates to the diameter PV, i.e. those which are parallel to QV.

---

* i.e. in the treatises on conics by Aristaeus and Euclid.

"For this is proved in the conics*."

Thus $QV^2 = p \cdot PV$.

And $BH^2 = p_a \cdot AH$, while $AH = PV$.

Therefore $QV^2 : BH^2 = p : p_a$.

But $QV^2 : QD^2 = p : p_a$;

hence $BH = QD$.

Thus $BH \cdot AH = QD \cdot PV$,

and therefore $\triangle ABB' = \triangle PQQ'$;

that is, the area of the triangle PQQ' is constant so long as PV is of constant length.

Hence also the area of the segment PQQ' is constant under the same conditions; for the segment is equal to $\frac{4}{3} \triangle PQQ'$. [*Quadrature of the Parabola*, Prop. 17 or 24.]

---

* The theorem which is here assumed by Archimedes as known can be proved in various ways.

(1) It is easily deduced from Apollonius I. 49 (cf. *Apollonius of Perga*, pp. liii, 39). If in the figure the tangents at A and P be drawn, the former meeting PV in E, and the latter meeting the axis in T, and if AE, PT meet at C, the proposition of Apollonius is to the effect that

$$CP : PE = p : 2PT,$$

where p is the parameter of the ordinates to PV.

(2) It may be proved independently as follows.

Let QQ' meet the axis in O, and let QM, $Q'M'$, PN be ordinates to the axis.

Then $AM : AM' = QM^2 : Q'M'^2 = OM^2 : OM'^2$,

whence $AM : MM' = OM^2 : OM^2 - OM'^2$

$= OM^2 : (OM - OM') \cdot MM'$,

so that $OM^2 = AM \cdot (OM - OM')$.

That is to say, $(AM - AO)^2 = AM \cdot (AM + AM' - 2AO)$,

or $AO^2 = AM \cdot AM'$.

And, since $QM^2 = p_a \cdot AM$, and $Q'M'^2 = p_a \cdot AM'$,

it follows that $QM \cdot Q'M' = p_a \cdot AO$ .... (α).

Now $QV^2 : QD^2 = QV^2 : \left(\dfrac{QM + Q'M'}{2}\right)^2$

$= QV^2 : \left(\dfrac{QM - Q'M'}{2}\right)^2 + QM \cdot Q'M'$

$= QV^2 : (PN^2 + QM \cdot Q'M')$

$= p \cdot PV : p_a \cdot (AN + AO)$, by (α).

But $PV = TO = AN + AO$.

Therefore $QV^2 : QD^2 = p : p_a$.

Proposition 4.

The area of any ellipse is to that of the auxiliary circle as the minor axis to the major.

Let AA' be the major and BB' the minor axis of the ellipse, and let BB' meet the auxiliary circle in b, b'.

Suppose O to be such a circle that

(circle $AbA'b'$) : $O = CA : CB$.

Then shall O be equal to the area of the ellipse.

For, if not, O must be either greater or less than the ellipse.

I. If possible, let O be greater than the ellipse.

We can then inscribe in the circle O an equilateral polygon of $4n$ sides such that its area is greater than that of the ellipse. [cf. *On the Sphere and Cylinder*, I. 6.]

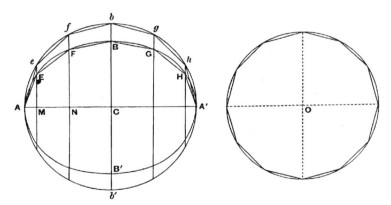

Let this be done, and inscribe in the auxiliary circle of the ellipse the polygon $AefbghA'...$ similar to that inscribed in O. Let the perpendiculars eM, fN,... on AA' meet the ellipse in E, F,... respectively. Join AE, EF, FB,....

Suppose that P' denotes the area of the polygon inscribed in the auxiliary circle, and P that of the polygon inscribed in the ellipse.

Then, since all the lines $eM, fN,...$ are cut in the same proportions at $E, F,...,$

i.e. $\qquad eM : EM = fN : FN = ... = bC : BC,$

the pairs of triangles, as $eAM, EAM,$ and the pairs of trapeziums, as $eMNf, EMNF,$ are all in the same ratio to one another as bC to BC, or as CA to CB.

Therefore, by addition,

$$P' : P = CA : CB.$$

Now P' : (polygon inscribed in O)

$\qquad\qquad\qquad =$ (circle $AbA'b'$) : O

$\qquad\qquad\qquad = CA : CB$, by hypothesis.

Therefore P is equal to the polygon inscribed in O.

But this is impossible, because the latter polygon is by hypothesis greater than the ellipse, and *a fortiori* greater than P.

Hence O is not greater than the ellipse.

II. If possible, let O be less than the ellipse.

In this case we inscribe in the *ellipse* a polygon P with $4n$ equal sides such that $P > O$.

Let the perpendiculars from the angular points on the axis AA' be produced to meet the auxiliary circle, and let the corresponding polygon (P') in the circle be formed.

Inscribe in O a polygon similar to P'.

Then $\qquad P' : P = CA : CB$

$\qquad\qquad\qquad =$ (circle $AbA'b'$) : O, by hypothesis,

$\qquad\qquad\qquad = P'$: (polygon inscribed in O).

Therefore the polygon inscribed in O is equal to the polygon P; which is impossible, because $P > O$.

Hence O, being neither greater nor less than the ellipse, is equal to it; and the required result follows.

Proposition 5.

If AA', BB' be the major and minor axis of an ellipse respectively, and if d be the diameter of any circle, then

$$(\text{area of ellipse}) : (\text{area of circle}) = AA' \cdot BB' : d^2.$$

For

(area of ellipse) : (area of auxiliary circle) $= BB' : AA'$ [Prop. 4]
$= AA' \cdot BB' : AA'^2.$

And

(area of aux. circle) : (area of circle with diam. d) $= AA'^2 : d^2$.

Therefore the required result follows *ex aequali*.

Proposition 6.

The areas of ellipses are as the rectangles under their axes.

This follows at once from Props. 4, 5.

COR. *The areas of similar ellipses are as the squares of corresponding axes.*

Proposition 7.

Given an ellipse with centre C, and a line CO drawn perpendicular to its plane, it is possible to find a circular cone with vertex O and such that the given ellipse is a section of it [*or, in other words, to find the circular sections of the cone with vertex O passing through the circumference of the ellipse*].

Conceive an ellipse with BB' as its minor axis and lying in a plane perpendicular to that of the paper. Let CO be drawn perpendicular to the plane of the ellipse, and let O be the vertex of the required cone. Produce OB, OC, OB', and in the same plane with them draw BED meeting OC, OB' produced in E, D respectively and in such a direction that

$$BE \cdot ED : EO^2 = CA^2 : CO^2,$$

where CA is half the major axis of the ellipse.

"And this is possible, since
$$BE.ED : EO^2 > BC.CB' : CO^2."$$
[Both the construction and this proposition are assumed as known.]

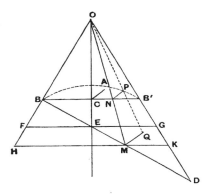

Now conceive a circle with BD as diameter lying in a plane at right angles to that of the paper, and describe a cone with this circle for its base and with vertex O.

We have therefore to prove that the given ellipse is a section of the cone, or, if P be any point on the ellipse, that P lies on the surface of the cone.

Draw PN perpendicular to BB'. Join ON and produce it to meet BD in M, and let MQ be drawn in the plane of the circle on BD as diameter perpendicular to BD and meeting the circle in Q. Also let FG, HK be drawn through E, M respectively parallel to BB'.

We have then
$$\begin{aligned}QM^2 : HM.MK &= BM.MD : HM.MK \\ &= BE.ED : FE.EG \\ &= (BE.ED : EO^2).(EO^2 : FE.EG) \\ &= (CA^2 : CO^2).(CO^2 : BC.CB') \\ &= CA^2 : CB^2 \\ &= PN^2 : BN.NB'.\end{aligned}$$

Therefore $QM^2 : PN^2 = HM . MK : BN . NB'$
$$= OM^2 : ON^2;$$
whence, since PN, QM are parallel, OPQ is a straight line.

But Q is on the circumference of the circle on BD as diameter; therefore OQ is a generator of the cone, and hence P lies on the cone.

Thus the cone passes through all points on the ellipse.

Proposition 8.

Given an ellipse, a plane through one of its axes AA' and perpendicular to the plane of the ellipse, and a line CO drawn from C, the centre, in the given plane through AA' but not perpendicular to AA', it is possible to find a cone with vertex O such that the given ellipse is a section of it [*or, in other words, to find the circular sections of the cone with vertex O whose surface passes through the circumference of the ellipse*].

By hypothesis, OA, OA' are unequal. Produce OA' to D so that $OA = OD$. Join AD, and draw FG through C parallel to it.

The given ellipse is to be supposed to lie in a plane perpendicular to the plane of the paper. Let BB' be the other axis of the ellipse.

Conceive a plane through AD perpendicular to the plane of the paper, and in it describe either (*a*), if $CB^2 = FC . CG$, a circle with diameter AD, or (*b*), if not, an ellipse on AD as axis such that, if d be the other axis,

$$d^2 : AD^2 = CB^2 : FC . CG.$$

Take a cone with vertex O whose surface passes through the circle or ellipse just drawn. This is possible even when the curve is an ellipse, because the line from O to the middle point of AD is perpendicular to the plane of the ellipse, and the construction is effected by means of Prop. 7.

Let P be any point on the given ellipse, and we have only to prove that P lies on the surface of the cone so described.

Draw PN perpendicular to AA'. Join ON, and produce it to meet AD in M. Through M draw HK parallel to $A'A$.

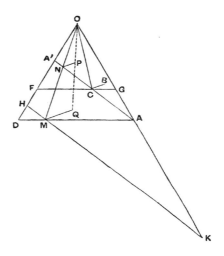

Lastly, draw MQ perpendicular to the plane of the paper (and therefore perpendicular to both HK and AD) meeting the ellipse or circle about AD (and therefore the surface of the cone) in Q.

Then
$$QM^2 : HM.MK = (QM^2 : DM.MA).(DM.MA : HM.MK)$$
$$= (d^2 : AD^2).(FC.CG : A'C.CA)$$
$$= (CB^2 : FC.CG).(FC.CG : A'C.CA)$$
$$= CB^2 : CA^2$$
$$= PN^2 : A'N.NA.$$

Therefore, alternately,
$$QM^2 : PN^2 = HM.MK : A'N.NA$$
$$= OM^2 : ON^2.$$

Thus, since PN, QM are parallel, OPQ is a straight line; and, Q being on the surface of the cone, it follows that P is also on the surface of the cone.

Similarly all points on the ellipse are also on the cone, and the ellipse is therefore a section of the cone.

Proposition 9.

Given an ellipse, a plane through one of its axes and perpendicular to that of the ellipse, and a straight line CO drawn from the centre C of the ellipse in the given plane through the axis but not perpendicular to that axis, it is possible to find a cylinder with axis OC such that the ellipse is a section of it [or, in other words, to find the circular sections of the cylinder with axis OC whose surface passes through the circumference of the given ellipse].

Let AA' be an axis of the ellipse, and suppose the plane of the ellipse to be perpendicular to that of the paper, so that OC lies in the plane of the paper.

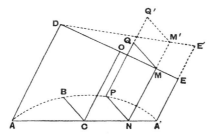

Draw AD, $A'E$ parallel to CO, and let DE be the line through O perpendicular to both AD and $A'E$.

We have now three different cases according as the other axis BB' of the ellipse is (1) equal to, (2) greater than, or (3) less than, DE.

(1) Suppose $BB' = DE$.

Draw a plane through DE at right angles to OC, and in this plane describe a circle on DE as diameter. Through this circle describe a cylinder with axis OC.

This cylinder shall be the cylinder required, or its surface shall pass through every point P of the ellipse.

For, if P be any point on the ellipse, draw PN perpendicular to AA'; through N draw NM parallel to CO meeting DE in M, and through M, in the plane of the circle on DE as diameter, draw MQ perpendicular to DE, meeting the circle in Q.

Then, since $DE = BB'$,
$$PN^2 : AN \cdot NA' = DO^2 : AC \cdot CA'.$$
And $DM \cdot ME : AN \cdot NA' = DO^2 : AC^2$,
since AD, NM, CO, $A'E$ are parallel.
Therefore
$$PN^2 = DM \cdot ME$$
$$= QM^2,$$
by the property of the circle.

Hence, since PN, QM are equal as well as parallel, PQ is parallel to MN and therefore to CO. It follows that PQ is a generator of the cylinder, whose surface accordingly passes through P.

(2) If $BB' > DE$, we take E' on $A'E$ such that $DE' = BB'$ and describe a circle on DE' as diameter in a plane perpendicular to that of the paper; and the rest of the construction and proof is exactly similar to those given for case (1).

(3) Suppose $BB' < DE$.

Take a point K on CO produced such that
$$DO^2 - CB^2 = OK^2.$$

From K draw KR perpendicular to the plane of the paper and equal to CB.

Thus $OR^2 = OK^2 + CB^2 = OD^2$.

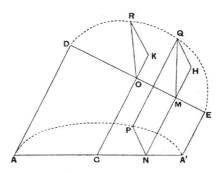

In the plane containing DE, OR describe a circle on DE as diameter. Through this circle (which must pass through R) draw a cylinder with axis OC.

ON CONOIDS AND SPHEROIDS. 121

We have then to prove that, if P be any point on the given ellipse, P lies on the cylinder so described.

Draw PN perpendicular to AA', and through N draw NM parallel to CO meeting DE in M. In the plane of the circle on DE as diameter draw MQ perpendicular to DE and meeting the circle in Q.

Lastly, draw QH perpendicular to NM produced. QH will then be perpendicular to the plane containing AC, DE, i.e. the plane of the paper.

Now $\quad QH^2 : QM^2 = KR^2 : OR^2$, by similar triangles.
And $\quad QM^2 : AN.NA' = DM.ME : AN.NA'$
$$= OD^2 : CA^2.$$
Hence, *ex aequali*, since $OR = OD$,
$$QH^2 : AN.NA' = KR^2 : CA^2$$
$$= CB^2 : CA^2$$
$$= PN^2 : AN.NA'.$$
Thus $QH = PN$. And QH, PN are also parallel. Accordingly PQ is parallel to MN, and therefore to CO, so that PQ is a generator, and the cylinder passes through P.

Proposition 10.

It was proved by the earlier geometers that *any two cones have to one another the ratio compounded of the ratios of their bases and of their heights**. The same method of proof will show that *any segments of cones have to one another the ratio compounded of the ratios of their bases and of their heights*.

The proposition that *any 'frustum' of a cylinder is triple of the conical segment which has the same base as the frustum and equal height* is also proved in the same manner as the proposition that *the cylinder is triple of the cone which has the same base as the cylinder and equal height*†.

---

* This follows from Eucl. xii. 11 and 14 taken together. Cf. *On the Sphere and Cylinder* i, Lemma 1.

† This proposition was proved by Eudoxus, as stated in the preface to *On the Sphere and Cylinder* i. Cf. Eucl. xii. 10.

Proposition 11.

(1) *If a paraboloid of revolution be cut by a plane through, or parallel to, the axis, the section will be a parabola equal to the original parabola which by its revolution generates the paraboloid. And the axis of the section will be the intersection between the cutting plane and the plane through the axis of the paraboloid at right angles to the cutting plane.*

If the paraboloid be cut by a plane at right angles to its axis, the section will be a circle whose centre is on the axis.

(2) *If a hyperboloid of revolution be cut by a plane through the axis, parallel to the axis, or through the centre, the section will be a hyperbola, (a) if the section be through the axis, equal, (b) if parallel to the axis, similar, (c) if through the centre, not similar, to the original hyperbola which by its revolution generates the hyperboloid. And the axis of the section will be the intersection of the cutting plane and the plane through the axis of the hyperboloid at right angles to the cutting plane.*

Any section of the hyperboloid by a plane at right angles to the axis will be a circle whose centre is on the axis.

(3) *If any of the spheroidal figures be cut by a plane through the axis or parallel to the axis, the section will be an ellipse, (a) if the section be through the axis, equal, (b) if parallel to the axis, similar, to the ellipse which by its revolution generates the figure. And the axis of the section will be the intersection of the cutting plane and the plane through the axis of the spheroid at right angles to the cutting plane.*

If the section be by a plane at right angles to the axis of the spheroid, it will be a circle whose centre is on the axis.

(4) *If any of the said figures be cut by a plane through the axis, and if a perpendicular be drawn to the plane of section from any point on the surface of the figure but not on the section, that perpendicular will fall within the section.*

" And the proofs of all these propositions are evident."*

* Cf. the Introduction, chapter III. § 4.

Proposition 12.

If a paraboloid of revolution be cut by a plane neither parallel nor perpendicular to the axis, and if the plane through the axis perpendicular to the cutting plane intersect it in a straight line of which the portion intercepted within the paraboloid is RR', the section of the paraboloid will be an ellipse whose major axis is RR' and whose minor axis is equal to the perpendicular distance between the lines through R, R' parallel to the axis of the paraboloid.

Suppose the cutting plane to be perpendicular to the plane of the paper, and let the latter be the plane through the axis ANF of the paraboloid which intersects the cutting plane at right angles in RR'. Let RH be parallel to the axis of the paraboloid, and $R'H$ perpendicular to RH.

Let Q be any point on the section made by the cutting plane, and from Q draw QM perpendicular to RR'. QM will therefore be perpendicular to the plane of the paper.

Through M draw $DMFE$ perpendicular to the axis ANF meeting the parabolic section made by the plane of the paper in D, E. Then QM is perpendicular to DE, and, if a plane be drawn through DE, QM, it will be perpendicular to the axis and will cut the paraboloid in a circular section.

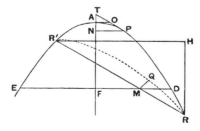

Since Q is on this circle,
$$QM^2 = DM \cdot ME.$$

Again, if PT be that tangent to the parabolic section in the plane of the paper which is parallel to RR', and if the tangent at A meet PT in O, then, from the property of the parabola,

$$DM \cdot ME : RM \cdot MR' = AO^2 : OP^2 \quad \text{[Prop. 3 (1)]}$$
$$= AO^2 : OT^2, \text{ since } AN = AT.$$

Therefore $\quad QM^2 : RM \cdot MR' = AO^2 : OT^2$
$$= R'H^2 : RR'^2,$$

by similar triangles.

Hence Q lies on an ellipse whose major axis is RR' and whose minor axis is equal to $R'H$.

Propositions 13, 14.

If a hyperboloid of revolution be cut by a plane meeting all the generators of the enveloping cone, or if an 'oblong' spheroid be cut by a plane not perpendicular to the axis*, and if a plane through the axis intersect the cutting plane at right angles in a straight line on which the hyperboloid or spheroid intercepts a length RR', then the section by the cutting plane will be an ellipse whose major axis is RR'.

Suppose the cutting plane to be at right angles to the plane of the paper, and suppose the latter plane to be that

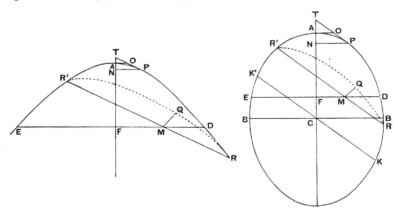

---

* Archimedes begins Prop. 14 for the *spheroid* with the remark that, when the cutting plane passes through or is parallel to the axis, the case is clear (δῆλον). Cf. Prop. 11 (3).

ON CONOIDS AND SPHEROIDS.

through the axis ANF which intersects the cutting plane at right angles in RR'. The section of the hyperboloid or spheroid by the plane of the paper is thus a hyperbola or ellipse having ANF for its transverse or major axis.

Take any point on the section made by the cutting plane, as Q, and draw QM perpendicular to RR'. QM will then be perpendicular to the plane of the paper.

Through M draw DFE at right angles to the axis ANF meeting the hyperbola or ellipse in D, E; and through QM, DE let a plane be described. This plane will accordingly be perpendicular to the axis and will cut the hyperboloid or spheroid in a circular section.

Thus $\qquad QM^2 = DM \cdot ME.$

Let PT be that tangent to the hyperbola or ellipse which is parallel to RR', and let the tangent at A meet PT in O.

Then, by the property of the hyperbola or ellipse,

$$DM \cdot ME : RM \cdot MR' = OA^2 : OP^2,$$
or $\qquad QM^2 : RM \cdot MR' = OA^2 : OP^2.$

Now (1) in the hyperbola $OA < OP$, because $AT < AN$*, and accordingly $OT < OP$, while $OA < OT$,

(2) in the ellipse, if KK' be the diameter parallel to RR', and BB' the minor axis,

$$BC \cdot CB' : KC \cdot CK' = OA^2 : OP^2;$$

and $\qquad BC \cdot CB' < KC \cdot CK'$, so that $OA < OP$.

Hence in both cases the locus of Q is an ellipse whose major axis is RR'.

COR. 1. If the spheroid be a 'flat' spheroid, the section will be an ellipse, and everything will proceed as before except that RR' will in this case be the *minor* axis.

COR. 2. In all conoids or spheroids parallel sections will be similar, since the ratio $OA^2 : OP^2$ is the same for all the parallel sections.

---

* With reference to this assumption cf. the Introduction, chapter III. § 3.

Proposition 15.

(1) *If from any point on the surface of a conoid a line be drawn, in the case of the paraboloid, parallel to the axis, and, in the case of the hyperboloid, parallel to any line passing through the vertex of the enveloping cone, the part of the straight line which is in the same direction as the convexity of the surface will fall without it, and the part which is in the other direction within it.*

For, if a plane be drawn, in the case of the paraboloid, through the axis and the point, and, in the case of the hyperboloid, through the given point and through the given straight line drawn through the vertex of the enveloping cone, the section by the plane will be (*a*) in the paraboloid a parabola whose axis is the axis of the paraboloid, (*b*) in the hyperboloid a hyperbola in which the given line through the vertex of the enveloping cone is a diameter*.   [Prop. 11]

Hence the property follows from the plane properties of the conics.

(2) *If a plane touch a conoid without cutting it, it will touch it at one point only, and the plane drawn through the point of contact and the axis of the conoid will be at right angles to the plane which touches it.*

For, if possible, let the plane touch at two points. Draw through each point a parallel to the axis. The plane passing through both parallels will therefore either pass through, or be parallel to, the axis. Hence the section of the conoid made by this plane will be a conic [Prop. 11 (1), (2)], the two points will lie on this conic, and the line joining them will lie within the conic and therefore within the conoid. But this line will be in the tangent plane, since the two points are in it. Therefore some portion of the tangent plane will be within the conoid; which is impossible, since the plane does not cut it.

---

* There seems to be some error in the text here, which says that "the *diameter*" (i.e. axis) of the hyperbola is "the straight line drawn in the conoid from the vertex of the cone." But this straight line is not, in general, the *axis* of the section.

Therefore the tangent plane touches in one point only.

That the plane through the point of contact and the axis is perpendicular to the tangent plane is evident in the particular case where the point of contact is the vertex of the conoid. For, if two planes through the axis cut it in two conics, the tangents at the vertex in both conics will be perpendicular to the axis of the conoid. And all such tangents will be in the tangent plane, which must therefore be perpendicular to the axis and to any plane through the axis.

If the point of contact P is not the vertex, draw the plane passing through the axis AN and the point P. It will cut the conoid in a conic whose axis is AN and the tangent plane in a line DPE touching the conic at P. Draw PNP' perpendicular to the axis, and draw a plane through it also perpendicular to the axis. This plane will make a circular section and meet the tangent plane in a tangent to the circle, which will therefore be at right angles to PN. Hence the tangent to the circle will be at right angles to the plane containing PN, AN; and it follows that this last plane is perpendicular to the tangent plane.

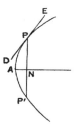

Proposition 16.

(1) *If a plane touch any of the spheroidal figures without cutting it, it will touch at one point only, and the plane through the point of contact and the axis will be at right angles to the tangent plane.*

This is proved by the same method as the last proposition.

(2) *If any conoid or spheroid be cut by a plane through the axis, and if through any tangent to the resulting conic a plane be erected at right angles to the plane of section, the plane so erected will touch the conoid or spheroid in the same point as that in which the line touches the conic.*

For it cannot meet the surface at any other point. If it did, the perpendicular from the second point on the cutting plane would be perpendicular also to the tangent to the conic and would therefore fall outside the surface. But it must fall within it. [Prop. 11 (4)]

(3) *If two parallel planes touch any of the spheroidal figures, the line joining the points of contact will pass through the centre of the spheroid.*

If the planes are at right angles to the axis, the proposition is obvious. If not, the plane through the axis and one point of contact is at right angles to the tangent plane at that point. It is therefore at right angles to the parallel tangent plane, and therefore passes through the second point of contact. Hence both points of contact lie on one plane through the axis, and the proposition is reduced to a plane one.

Proposition 17.

If two parallel planes touch any of the spheroidal figures, and another plane be drawn parallel to the tangent planes and passing through the centre, the line drawn through any point of the circumference of the resulting section parallel to the chord of contact of the tangent planes will fall outside the spheroid.

This is proved at once by reduction to a plane proposition.

Archimedes adds that it is evident that, if the plane parallel to the tangent planes does not pass through the centre, a straight line drawn in the manner described will fall without the spheroid in the direction of the smaller segment but within it in the other direction.

Proposition 18.

Any spheroidal figure which is cut by a plane through the centre is divided, both as regards its surface and its volume, into two equal parts by that plane.

To prove this, Archimedes takes another equal and similar spheroid, divides it similarly by a plane through the centre, and then uses the method of application.

Propositions 19, 20.

Given a segment cut off by a plane from a paraboloid or hyperboloid of revolution, or a segment of a spheroid less than half the spheroid also cut off by a plane, it is possible to inscribe in the segment one solid figure and to circumscribe about it another solid figure, each made up of cylinders or 'frusta' of cylinders of equal height, and such that the circumscribed figure exceeds the inscribed figure by a volume less than that of any given solid.

Let the plane base of the segment be perpendicular to the plane of the paper, and let the plane of the paper be the plane through the axis of the conoid or spheroid which cuts the base of the segment at right angles in BC. The section in the plane of the paper is then a conic BAC. [Prop. 11]

Let EAF be that tangent to the conic which is parallel to BC, and let A be the point of contact. Through EAF draw a plane parallel to the plane through BC bounding the segment. The plane so drawn will then touch the conoid or spheroid at A. [Prop. 16]

(1) If the base of the segment is at right angles to the axis of the conoid or spheroid, A will be the vertex of the conoid or spheroid, and its axis AD will bisect BC at right angles.

(2) If the base of the segment is not at right angles to the axis of the conoid or spheroid, we draw AD

(a) in the paraboloid, parallel to the axis,

(b) in the hyperboloid, through the centre (or the vertex of the enveloping cone),

(c) in the spheroid, through the centre,

and in all the cases it will follow that AD bisects BC in D.

Then A will be the vertex of the segment, and AD will be its axis.

Further, the base of the segment will be a circle or an ellipse with BC as diameter or as an axis respectively, and with centre D. We can therefore describe through this circle

or ellipse a cylinder or a 'frustum' of a cylinder whose axis is
AD. [Prop. 9]

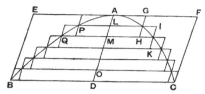

Dividing this cylinder or frustum continually into equal parts by planes parallel to the base, we shall at length arrive at a cylinder or frustum less in volume than any given solid.

Let this cylinder or frustum be that whose axis is OD, and let AD be divided into parts equal to OD, at $L, M,....$ Through $L, M,...$ draw lines parallel to BC meeting the conic in $P, Q,...$, and through these lines draw planes parallel to the base of the segment. These will cut the conoid or spheroid in circles or similar ellipses. On each of these circles or ellipses describe two cylinders or frusta of cylinders each with axis equal to OD, one of them lying in the direction of A and the other in the direction of D, as shown in the figure.

Then the cylinders or frusta of cylinders drawn in the direction of A make up a circumscribed figure, and those in the direction of D an inscribed figure, in relation to the segment.

Also the cylinder or frustum PG in the circumscribed figure is equal to the cylinder or frustum PH in the inscribed figure, QI in the circumscribed figure is equal to QK in the inscribed figure, and so on.

Therefore, by addition,

(circumscribed fig.) = (inscr. fig.)
+ (cylinder or frustum whose axis is OD).

But the cylinder or frustum whose axis is OD is less than the given solid figure; whence the proposition follows.

"Having set out these preliminary propositions, let us proceed to demonstrate the theorems propounded with reference to the figures."

Propositions **21, 22**.

Any segment of a paraboloid of revolution is half as large again as the cone or segment of a cone which has the same base and the same axis.

Let the base of the segment be perpendicular to the plane of the paper, and let the plane of the paper be the plane through the axis of the paraboloid which cuts the base of the segment at right angles in BC and makes the parabolic section BAC.

Let EF be that tangent to the parabola which is parallel to BC, and let A be the point of contact.

Then (1), if the plane of the base of the segment is perpendicular to the axis of the paraboloid, that axis is the line AD bisecting BC at right angles in D.

(2) If the plane of the base is not perpendicular to the axis of the paraboloid, draw AD parallel to the axis of the paraboloid. AD will then bisect BC, but not at right angles.

Draw through EF a plane parallel to the base of the segment. This will touch the paraboloid at A, and A will be the vertex of the segment, AD its axis.

The base of the segment will be a circle with diameter BC or an ellipse with BC as major axis.

Accordingly a cylinder or a frustum of a cylinder can be found passing through the circle or ellipse and having AD for its axis [Prop. 9]; and likewise a cone or a segment of a cone can be drawn passing through the circle or ellipse and having A for vertex and AD for axis. [Prop. 8]

Suppose X to be a cone equal to $\frac{3}{2}$ (cone or segment of cone ABC). The cone X is therefore equal to half the cylinder or frustum of a cylinder EC. [Cf. Prop. 10]

We shall prove that the volume of the segment of the paraboloid is equal to X.

If not, the segment must be either greater or less than X.

I. If possible, let the segment be greater than X.

We can then inscribe and circumscribe, as in the last

proposition, figures made up of cylinders or frusta of cylinders with equal height and such that

(circumscribed fig.) − (inscribed fig.) < (segment) − X.

Let the greatest of the cylinders or frusta forming the circumscribed figure be that whose base is the circle or ellipse about BC and whose axis is OD, and let the smallest of them be that whose base is the circle or ellipse about PP' and whose axis is AL.

Let the greatest of the cylinders forming the inscribed figure be that whose base is the circle or ellipse about RR' and whose axis is OD, and let the smallest be that whose base is the circle or ellipse about PP' and whose axis is LM.

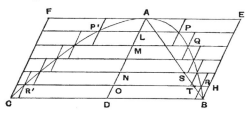

Produce all the plane bases of the cylinders or frusta to meet the surface of the complete cylinder or frustum EC.

Now, since

(circumscribed fig.) − (inscr. fig.) < (segment) − X,

it follows that   (inscribed figure) > X ....(α).

Next, comparing successively the cylinders or frusta with heights equal to OD and respectively forming parts of the complete cylinder or frustum EC and of the inscribed figure, we have

(first cylinder or frustum in EC) : (first in inscr. fig.)

$= BD^2 : RO^2$

$= AD : AO$

$= BD : TO$, where AB meets OR in T.

And (second cylinder or frustum in EC) : (second in inscr. fig.)

$= HO : SN$, in like manner,

and so on.

Hence [Prop. 1] (cylinder or frustum EC) : (inscribed figure)
$$= (BD + HO + \ldots) : (TO + SN + \ldots),$$
where $BD, HO, \ldots$ are all equal, and $BD, TO, SN, \ldots$ diminish in arithmetical progression.

But [Lemma preceding Prop. 1]
$$BD + HO + \ldots > 2(TO + SN + \ldots).$$
Therefore (cylinder or frustum EC) > 2 (inscribed fig.),
or X > (inscribed fig.);
which is impossible, by (α) above.

II. If possible, let the segment be less than X.

In this case we inscribe and circumscribe figures as before, but such that

(circumscr. fig.) − (inscr. fig.) < X − (segment),
whence it follows that
$$\text{(circumscribed figure)} < X \quad \ldots\ldots\ldots\ldots\ldots(\beta).$$

And, comparing the cylinders or frusta making up the complete cylinder or frustum CE and the *circumscribed* figure respectively, we have

(first cylinder or frustum in CE) : (first in circumscr. fig.)
$$= BD^2 : BD^2$$
$$= BD : BD.$$
(second in CE) : (second in circumscr. fig.)
$$= HO^2 : RO^2$$
$$= AD : AO$$
$$= HO : TO,$$
and so on.

Hence [Prop. 1]
(cylinder or frustum CE) : (circumscribed fig.)
$$= (BD + HO + \ldots) : (BD + TO + \ldots),$$
$$< 2 : 1, \quad \text{[Lemma preceding Prop. 1]}$$
and it follows that
$$X < \text{(circumscribed fig.)};$$
which is impossible, by (β).

Thus the segment, being neither greater nor less than X, is equal to it, and therefore to $\frac{3}{2}$ (cone or segment of cone ABC).

134   ARCHIMEDES

Proposition 23.

If from a paraboloid of revolution two segments be cut off, one by a plane perpendicular to the axis, the other by a plane not perpendicular to the axis, and if the axes of the segments are equal, the segments will be equal in volume.

Let the two planes be supposed perpendicular to the plane of the paper, and let the latter plane be the plane through the axis of the paraboloid cutting the other two planes at right angles in BB', QQ' respectively and the paraboloid itself in the parabola $QPQ'B'$.

Let AN, PV be the equal axes of the segments, and A, P their respective vertices.

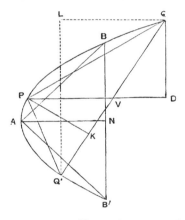

Draw QL parallel to AN or PV and $Q'L$ perpendicular to QL.

Now, since the segments of the parabolic section cut off by BB', QQ' have equal axes, the triangles ABB', PQQ' are equal [Prop. 3]. Also, if QD be perpendicular to PV, $QD = BN$ (as in the same Prop. 3).

Conceive two cones drawn with the same bases as the segments and with A, P as vertices respectively. The height of the cone PQQ' is then PK, where PK is perpendicular to QQ'.

Now the cones are in the ratio compounded of the ratios of their bases and of their heights, i.e. the ratio compounded of (1) the ratio of the circle about BB' to the ellipse about QQ', and (2) the ratio of AN to PK.

That is to say, we have, by means of Props. 5, 12,
(cone ABB') : (cone PQQ') = $(BB'^2 : QQ' . Q'L).(AN : PK)$.
And $BB' = 2BN = 2QD = Q'L$, while $QQ' = 2QV$.
Therefore
(cone ABB') : (cone PQQ') = $(QD : QV).(AN : PK)$
$= (PK : PV).(AN : PK)$
$= AN : PV$.

Since $AN = PV$, the ratio of the cones is a ratio of equality; and it follows that the segments, being each half as large again as the respective cones [Prop. 22], are equal.

Proposition 24.

If from a paraboloid of revolution two segments be cut off by planes drawn in any manner, the segments will be to one another as the squares on their axes.

For let the paraboloid be cut by a plane through the axis in the parabolic section $P'PApp'$, and let the axis of the parabola and paraboloid be ANN'.

Measure along ANN' the lengths AN, AN' equal to the respective axes of the given segments, and through N, N' draw planes perpendicular to the axis, making circular sections on Pp, $P'p'$ as diameters respectively. With these circles as bases and with the common vertex A let two cones be described.

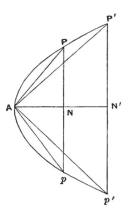

Now the segments of the paraboloid whose bases are the circles about Pp, $P'p'$ are equal to the given segments respectively, since their respective axes are equal [Prop. 23]; and, since the segments APp, $AP'p'$ are half as large again as the cones APp, $AP'p'$ respectively, we have only to show that the cones are in the ratio of AN^2 to AN'^2.

But

$$(\text{cone } APp) : (\text{cone } AP'p') = (PN^2 : P'N'^2).(AN : AN')$$
$$= (AN : AN').(AN : AN')$$
$$= AN^2 : AN'^2;$$

thus the proposition is proved.

Propositions 25, 26.

In any hyperboloid of revolution, if A be the vertex and AD the axis of any segment cut off by a plane, and if CA be the semidiameter of the hyperboloid through A (CA being of course in the same straight line with AD), then

(segment) : (cone with same base and axis)

$$= (AD + 3CA) : (AD + 2CA).$$

Let the plane cutting off the segment be perpendicular to the plane of the paper, and let the latter plane be the plane through the axis of the hyperboloid which intersects the cutting plane at right angles in BB', and makes the hyperbolic segment BAB'. Let C be the centre of the hyperboloid (or the vertex of the enveloping cone).

Let EF be that tangent to the hyperbolic section which is parallel to BB'. Let EF touch at A, and join CA. Then CA produced will bisect BB' at D, CA will be a semi-diameter of the hyperboloid, A will be the vertex of the segment, and AD its axis. Produce AC to A' and H, so that $AC = CA' = A'H$.

Through EF draw a plane parallel to the base of the segment. This plane will touch the hyperboloid at A.

Then (1), if the base of the segment is at right angles to the axis of the hyperboloid, A will be the vertex, and AD the axis, of the hyperboloid as well as of the segment, and the base of the segment will be a circle on BB' as diameter.

ON CONOIDS AND SPHEROIDS. 137

(2) *If the base of the segment is not perpendicular to the axis of the hyperboloid, the base will be an ellipse on BB' as major axis.* [Prop. 13]

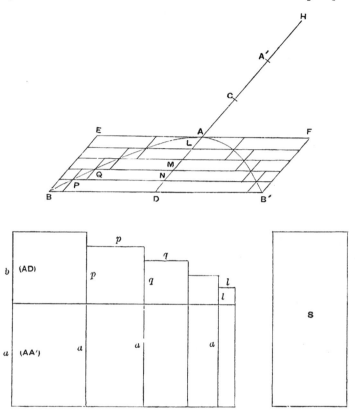

Then we can draw a cylinder or a frustum of a cylinder $EBB'F$ passing through the circle or ellipse about BB' and having AD for its axis; also we can describe a cone or a segment of a cone through the circle or ellipse and having A for its vertex.

We have to prove that

(segment ABB') : (cone or segment of cone ABB') = $HD : A'D$.

Let V be a cone such that

V : (cone or segment of cone ABB') $= HD : A'D$,......(α)

and we have to prove that V is equal to the segment.

Now

(cylinder or frustum EB') : (cone or segmt. of cone ABB') $= 3 : 1$.

Therefore, by means of (α),

(cylinder or frustum EB') : $V = A'D : \dfrac{HD}{3}$ ......(β).

If the segment is not equal to V, it must either be greater or less.

I. If possible, let the segment be greater than V.

Inscribe and circumscribe to the segment figures made up of cylinders or frusta of cylinders, with axes along AD and all equal to one another, such that

(circumscribed fig.) $-$ (inscr. fig.) $<$ (segmt.) $- V$,

whence (inscribed figure) $> V$ ....(γ).

Produce all the planes forming the bases of the cylinders or frusta of cylinders to meet the surface of the complete cylinder or frustum EB'.

Then, if ND be the axis of the greatest cylinder or frustum in the circumscribed figure, the complete cylinder will be divided into cylinders or frusta each equal to this greatest cylinder or frustum.

Let there be a number of straight lines a equal to AA' and as many in number as the parts into which AD is divided by the bases of the cylinders or frusta. To each line a apply a rectangle which shall overlap it by a square, and let the greatest of the rectangles be equal to the rectangle $AD . A'D$ and the least equal to the rectangle $AL . A'L$; also let the sides of the overlapping squares $b, p, q, ... l$ be in descending arithmetical progression. Thus $b, p, q, ... l$ will be respectively equal to AD, $AN, AM, ... AL$, and the rectangles $(ab + b^2), (ap + p^2), ... (al + l^2)$ will be respectively equal to $AD . A'D, AN . A'N, ... AL . A'L$.

Suppose, further, that we have a series of spaces S each equal to the largest rectangle AD. $A'D$ and as many in number as the diminishing rectangles.

Comparing now the successive cylinders or frusta (1) in the complete cylinder or frustum EB' and (2) in the inscribed figure, beginning from the base of the segment, we have

(first cylinder or frustum in EB') : (first in inscr. figure)
$$= BD^2 : PN^2$$
$$= AD \cdot A'D : AN \cdot A'N, \text{ from the hyperbola,}$$
$$= S : (ap + p^2).$$

Again

(second cylinder or frustum in EB') : (second in inscr. fig.)
$$= BD^2 : QM^2$$
$$= AD \cdot A'D : AM \cdot A'M$$
$$= S : (aq + q^2),$$

and so on.

The last cylinder or frustum in the complete cylinder or frustum EB' has no cylinder or frustum corresponding to it in the inscribed figure.

Combining the proportions, we have [Prop. 1]

(cylinder or frustum EB') : (inscribed figure)

$$= \text{(sum of all the spaces } S) : (ap + p^2) + (aq + q^2) + \ldots$$

$$> (a+b) : \left(\frac{a}{2} + \frac{b}{3}\right) \qquad \text{[Prop. 2]}$$

$$> A'D : \frac{HD}{3}, \quad \text{since } a = AA', \quad b = AD,$$

$$> (EB') : V, \quad \text{by } (\beta) \text{ above.}$$

Hence (inscribed figure) $< V$.

But this is impossible, because, by (γ) above, the inscribed figure is greater than V.

II. Next suppose, if possible, that the segment is less than V.

In this case we circumscribe and inscribe figures such that

(circumscribed fig.) − (inscribed fig.) < V − (segment),

whence we derive

V > (circumscribed figure) ....(δ).

We now compare successive cylinders or frusta in the complete cylinder or frustum and in the *circumscribed* figure; and we have

(first cylinder or frustum in EB') : (first in circumscribed fig.)

$$= S : S$$
$$= S : (ab + b^2),$$

(second in EB') : (second in circumscribed fig.)

$$= S : (ap + p^2),$$

and so on.

Hence [Prop. 1]

(cylinder or frustum EB') : (circumscribed fig.)

$$= \text{(sum of all spaces } S) : (ab + b^2) + (ap + p^2) + \dots$$

$$< (a+b) : \left(\frac{a}{2} + \frac{b}{3}\right) \qquad [\text{Prop. 2}]$$

$$< A'D : \frac{HD}{3}$$

$$< (EB') : V, \text{ by } (\beta) \text{ above.}$$

Hence the circumscribed figure is greater than V; which is impossible, by (δ) above.

Thus the segment is neither greater nor less than V, and is therefore equal to it.

Therefore, by (α),

(segment ABB') : (cone or segment of cone ABB')
$$= (AD + 3CA) : (AD + 2CA).$$

Propositions 27, 28, 29, 30.

(1) *In any spheroid whose centre is C, if a plane meeting the axis cut off a segment not greater than half the spheroid and having A for its vertex and AD for its axis, and if A'D be the axis of the remaining segment of the spheroid, then*

(*first segmt.*) : (*cone or segmt. of cone with same base and axis*)

$$= CA + A'D : A'D$$
$$[= 3CA - AD : 2CA - AD].$$

(2) *As a particular case, if the plane passes through the centre, so that the segment is half the spheroid, half the spheroid is double of the cone or segment of a cone which has the same vertex and axis.*

Let the plane cutting off the segment be at right angles to the plane of the paper, and let the latter plane be the plane through the axis of the spheroid which intersects the cutting plane in BB' and makes the elliptic section $ABA'B'$.

Let EF, $E'F'$ be the two tangents to the ellipse which are parallel to BB', let them touch it in A, A', and through the tangents draw planes parallel to the base of the segment. These planes will touch the spheroid at A, A', which will be the vertices of the two segments into which it is divided. Also AA' will pass through the centre C and bisect BB' in D.

Then (1) if the base of the segments be perpendicular to the axis of the spheroid, A, A' will be the vertices of the spheroid as well as of the segments, AA' will be the axis of the spheroid, and the base of the segments will be a circle on BB' as diameter;

(2) if the base of the segments be not perpendicular to the axis of the spheroid, the base of the segments will be an ellipse of which BB' is one axis, and AD, $A'D$ will be the axes of the segments respectively.

We can now draw a cylinder or a frustum of a cylinder $EBB'F$ through the circle or ellipse about BB' and having AD for its axis; and we can also draw a cone or a segment of a cone passing through the circle or ellipse about BB' and having A for its vertex.

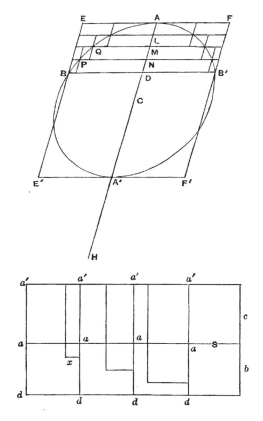

We have then to show that, if CA' be produced to H so that $CA' = A'H$,

(segment ABB') : (cone or segment of cone ABB') $= HD : A'D$.

Let V be such a cone that

V : (cone or segment of cone ABB') $= HD : A'D$ ... (α);

and we have to show that the segment ABB' is equal to V.

ON CONOIDS AND SPHEROIDS.   143

But, since

(cylinder or frustum EB') : (cone or segment of cone ABB')
$$= 3 : 1,$$
we have, by the aid of (α),

(cylinder or frustum EB') : $V = A'D : \dfrac{HD}{3}$ ......(β).

Now, if the segment ABB' is not equal to V, it must be either greater or less.

I. Suppose, if possible, that the segment is greater than V.

Let figures be inscribed and circumscribed to the segment consisting of cylinders or frusta of cylinders, with axes along AD and all equal to one another, such that

(circumscribed fig.) − (inscribed fig.) < (segment) − V,
whence it follows that

(inscribed fig.) > V ....(γ).

Produce all the planes forming the bases of the cylinders or frusta to meet the surface of the complete cylinder or frustum EB'. Thus, if ND be the axis of the greatest cylinder or frustum of a cylinder in the circumscribed figure, the complete cylinder or frustum EB' will be divided into cylinders or frusta of cylinders each equal to the greatest of those in the circumscribed figure.

Take straight lines da' each equal to $A'D$ and as many in number as the parts into which AD is divided by the bases of the cylinders or frusta, and measure da along da' equal to AD. It follows that $aa' = 2CD$.

Apply to each of the lines $a'd$ rectangles with height equal to ad, and draw the squares on each of the lines ad as in the figure. Let S denote the area of each complete rectangle.

From the first rectangle take away a gnomon with breadth equal to AN (i.e. with each end of a length equal to AN); take away from the second rectangle a gnomon with breadth equal to AM, and so on, the last rectangle having no gnomon taken from it.

Then

the first gnomon $= A'D \cdot AD - ND \cdot (A'D - AN)$
$= A'D \cdot AN + ND \cdot AN$
$= AN \cdot A'N.$

Similarly,

the second gnomon $= AM \cdot A'M,$

and so on.

And the last gnomon (that in the last rectangle but one) is equal to $AL \cdot A'L.$

Also, after the gnomons are taken away from the successive rectangles, the remainders (which we will call $R_1, R_2, \ldots R_n$, where n is the number of rectangles and accordingly $R_n = S$) are rectangles applied to straight lines each of length aa' and "exceeding by squares" whose sides are respectively equal to $DN, DM, \ldots DA.$

For brevity, let DN be denoted by x, and aa' or $2CD$ by c, so that $R_1 = cx + x^2$, $R_2 = c \cdot 2x + (2x)^2, \ldots$

Then, comparing successively the cylinders or frusta of cylinders (1) in the complete cylinder or frustum EB' and (2) in the inscribed figure, we have

(first cylinder or frustum in EB') : (first in inscribed fig.)

$= BD^2 : PN^2$

$= AD \cdot A'D : AN \cdot A'N$

$= S :$ (first gnomon) ;

(second cylinder or frustum in EB') : (second in inscribed fig.)

$= S :$ (second gnomon),

and so on.

The last of the cylinders or frusta in the cylinder or frustum EB' has none corresponding to it in the inscribed figure, and there is no corresponding gnomon.

Combining the proportions, we have [by Prop. 1]

(cylinder or frustum EB') : (inscribed fig.)

$=$ (sum of all spaces S) : (sum of gnomons).

Now the differences between S and the successive gnomons are $R_1, R_2, \ldots R_n$, while
$$R_1 = cx + x^2,$$
$$R_2 = c \cdot 2x + (2x)^2,$$
$$\ldots\ldots\ldots\ldots\ldots\ldots\ldots$$
$$R_n = cb + b^2 = S,$$
where $b = nx = AD$.

Hence [Prop. 2]

(sum of all spaces S) : $(R_1 + R_2 + \ldots + R_n) < (c + b) : \left(\dfrac{c}{2} + \dfrac{b}{3}\right)$.

It follows that

(sum of all spaces S) : (sum of gnomons) $> (c + b) : \left(\dfrac{c}{2} + \dfrac{2b}{3}\right)$
$$> A'D : \dfrac{HD}{3}.$$

Thus (cylinder or frustum EB') : (inscribed fig.)
$$> A'D : \dfrac{HD}{3}$$
$$> \text{(cylinder or frustum } EB') : V,$$
from (β) above.

Therefore (inscribed fig.) $< V$;

which is impossible, by (γ) above.

Hence the segment ABB' is not greater than V.

II. If possible, let the segment ABB' be less than V.

We then inscribe and circumscribe figures such that

(circumscribed fig.) − (inscribed fig.) $< V − $ (segment),

whence $V >$ (circumscribed fig.)................(δ).

In this case we compare the cylinders or frusta in (EB') with those in the *circumscribed* figure.

Thus

(first cylinder or frustum in EB') : (first in circumscribed fig.)
$$= S : S;$$
(second in EB') : (second in circumscribed fig.)
$$= S : \text{(first gnomon)},$$
and so on.

Lastly (last in EB') : (last in circumscribed fig.)
$$= S : \text{(last gnomon)}.$$

Now
$$\{S + \text{(all the gnomons)}\} = nS - (R_1 + R_2 + \ldots + R_{n-1}).$$

And $nS : R_1 + R_2 + \ldots + R_{n-1} > (c+b) : \left(\dfrac{c}{2} + \dfrac{b}{3}\right)$, [Prop. 2]

so that
$$nS : \{S + \text{(all the gnomons)}\} < (c+b) : \left(\dfrac{c}{2} + \dfrac{2b}{3}\right).$$

It follows that, if we combine the above proportions as in Prop. 1, we obtain

(cylinder or frustum EB') : (circumscribed fig.)
$$< (c+b) : \left(\dfrac{c}{2} + \dfrac{2b}{3}\right)$$
$$< A'D : \dfrac{HD}{3}$$
$$< (EB') : V, \text{ by } (\beta) \text{ above.}$$

Hence the circumscribed figure is greater than V; which is impossible, by (δ) above.

Thus, since the segment ABB' is neither greater nor less than V, it is equal to it; and the proposition is proved.

(2) The particular case [Props. 27, 28] where the segment is half the spheroid differs from the above in that the distance CD or $c/2$ vanishes, and the rectangles $cb + b^2$ are simply squares (b^2), so that the gnomons are simply the differences between b^2 and x^2, b^2 and $(2x)^2$, and so on.

Instead therefore of Prop. 2 we use the *Lemma to Prop. 2, Cor.* 1, given above [*On Spirals,* Prop. 10], and instead of the ratio $(c+b) : \left(\dfrac{c}{2} + \dfrac{2b}{3}\right)$ we obtain the ratio $3 : 2$, whence

(segment ABB') : (cone or segment of cone ABB') $= 2 : 1$.

[This result can also be obtained by simply substituting CA for AD in the ratio $(3CA - AD) : (2CA - AD)$.]

Propositions 31, 32.

If a plane divide a spheroid into two unequal segments, and if AN, A'N be the axes of the lesser and greater segments respectively, while C is the centre of the spheroid, then

(greater segmt.) : (cone or segmt. of cone with same base and axis)
$$= CA + AN : AN.$$

Let the plane dividing the spheroid be that through PP' perpendicular to the plane of the paper, and let the latter plane be that through the axis of the spheroid which intersects the cutting plane in PP' and makes the elliptic section $PAP'A'$.

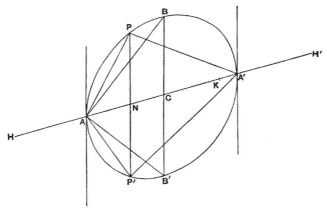

Draw the tangents to the ellipse which are parallel to PP'; let them touch the ellipse at A, A', and through the tangents draw planes parallel to the base of the segments. These planes will touch the spheroid at A, A', the line AA' will pass through the centre C and bisect PP' in N, while $AN, A'N$ will be the axes of the segments.

Then (1) if the cutting plane be perpendicular to the axis of the spheroid, AA' will be that axis, and A, A' will be the vertices of the spheroid as well as of the segments. Also the sections of the spheroid by the cutting plane and all planes parallel to it will be circles.

(2) If the cutting plane be not perpendicular to the axis,

the base of the segments will be an ellipse of which PP' is an axis, and the sections of the spheroid by all planes parallel to the cutting plane will be similar ellipses.

Draw a plane through C parallel to the base of the segments and meeting the plane of the paper in BB'.

Construct three cones or segments of cones, two having A for their common vertex and the plane sections through PP', BB' for their respective bases, and a third having the plane section through PP' for its base and A' for its vertex.

Produce CA to H and CA' to H' so that
$$AH = A'H' = CA.$$

We have then to prove that

(segment $A'PP'$) : (cone or segment of cone $A'PP'$)
$$= CA + AN : AN$$
$$= NH : AN.$$

Now half the spheroid is double of the cone or segment of a cone ABB' [Props. 27, 28]. Therefore

(the spheroid) = 4 (cone or segment of cone ABB').

But

(cone or segmt. of cone ABB') : (cone or segmt. of cone APP')
$$= (CA : AN).(BC^2 : PN^2)$$
$$= (CA : AN).(CA.CA' : AN.A'N)\ldots(\alpha).$$

If we measure AK along AA' so that
$$AK : AC = AC : AN,$$
we have $\qquad AK.A'N : AC.A'N = CA : AN,$

and the compound ratio in (α) becomes
$$(AK.A'N : CA.A'N).(CA.CA' : AN.A'N),$$
i.e. $\qquad AK.CA' : AN.A'N.$

Thus

(cone or segmt. of cone ABB') : (cone or segmt. of cone APP')
$$= AK.CA' : AN.A'N.$$

ON CONOIDS AND SPHEROIDS. 149

But   (cone or segment of cone APP') : (segment APP')
$$= A'N : NH' \quad \text{[Props. 29, 30]}$$
$$= AN.A'N : AN.NH'.$$

Therefore, *ex aequali*,

(cone or segment of cone ABB') : (segment APP')
$$= AK.CA' : AN.NH',$$

so that   (spheroid) : (segment APP')
$$= HH'.AK : AN.NH',$$

since $HH' = 4CA'.$

Hence   (segment $A'PP'$) : (segment APP')
$$= (HH'.AK - AN.NH') : AN.NH'$$
$$= (AK.NH + NH'.NK) : AN.NH'.$$

Further,

(segment APP') : (cone or segment of cone APP')
$$= NH' : A'N$$
$$= AN.NH' : AN.A'N,$$

and

(cone or segmt. of cone APP') : (cone or segmt. of cone $A'PP'$)
$$= AN : A'N$$
$$= AN.A'N : A'N^2.$$

From the last three proportions we obtain, *ex aequali*,

(segment $A'PP'$) : (cone or segment of cone $A'PP'$)
$$= (AK.NH + NH'.NK) : A'N^2$$
$$= (AK.NH + NH'.NK) : (CA^2 + NH'.CN)$$
$$= (AK.NH + NH'.NK) : (AK.AN + NH'.CN)\ldots(\beta).$$

But
$$AK.NH : AK.AN = NH : AN$$
$$= CA + AN : AN$$
$$= AK + CA : CA$$
$$\qquad\qquad\text{(since } AK : AC = AC : AN)$$
$$= HK : CA$$
$$= HK - NH : CA - AN$$
$$= NK : CN$$
$$= NH'.NK : NH'.CN.$$

Hence the ratio in (β) is equal to the ratio
$$AK.NH : AK.AN, \text{ or } NH : AN.$$

Therefore

(segment $A'PP'$) : (cone or segment of cone $A'PP'$)
$$= NH : AN$$
$$= CA + AN : AN.$$

[If (x, y) be the coordinates of P referred to the conjugate diameters AA', BB' as axes of x, y, and if $2a$, $2b$ be the lengths of the diameters respectively, we have, since

(spheroid) − (lesser segment) = (greater segment),

$$4.ab^2 - \frac{2a+x}{a+x} \cdot y^2(a-x) = \frac{2a-x}{a-x} \cdot y^2(a+x);$$

and the above proposition is the geometrical proof of the truth of this equation where x, y are connected by the equation

$$\frac{x^2}{a^2} + \frac{y^2}{b^2} = 1.]$$

ON SPIRALS.

"ARCHIMEDES to Dositheus greeting.

Of most of the theorems which I sent to Conon, and of which you ask me from time to time to send you the proofs, the demonstrations are already before you in the books brought to you by Heracleides; and some more are also contained in that which I now send you. Do not be surprised at my taking a considerable time before publishing these proofs. This has been owing to my desire to communicate them first to persons engaged in mathematical studies and anxious to investigate them. In fact, how many theorems in geometry which have seemed at first impracticable are in time successfully worked out! Now Conon died before he had sufficient time to investigate the theorems referred to; otherwise he would have discovered and made manifest all these things, and would have enriched geometry by many other discoveries besides. For I know well that it was no common ability that he brought to bear on mathematics, and that his industry was extraordinary. But, though many years have elapsed since Conon's death, I do not find that any one of the problems has been stirred by a single person. I wish now to put them in review one by one, particularly as it happens that there are two included among them which are impossible of realisation* [and which may serve as a warning] how those who claim to discover everything but produce no proofs of the same may be confuted as having actually pretended to discover the impossible.

---

* Heiberg reads τέλος δὲ ποθεσόμενα, but F has τέλους, so that the true reading is perhaps τέλους δὲ ποτιδεόμενα. The meaning appears to be simply 'wrong.'

What are the problems I mean, and what are those of which you have already received the proofs, and those of which the proofs are contained in this book respectively, I think it proper to specify. The first of the problems was, Given a sphere, to find a plane area equal to the surface of the sphere; and this was first made manifest on the publication of the book concerning the sphere, for, when it is once proved that the surface of any sphere is four times the greatest circle in the sphere, it is clear that it is possible to find a plane area equal to the surface of the sphere. The second was, Given a cone or a cylinder, to find a sphere equal to the cone or cylinder; the third, To cut a given sphere by a plane so that the segments of it have to one another an assigned ratio; the fourth, To cut a given sphere by a plane so that the segments of the surface have to one another an assigned ratio; the fifth, To make a given segment of a sphere similar to a given segment of a sphere*; the sixth, Given two segments of either the same or different spheres, to find a segment of a sphere which shall be similar to one of the segments and have its surface equal to the surface of the other segment. The seventh was, From a given sphere to cut off a segment by a plane so that the segment bears to the cone which has the same base as the segment and equal height an assigned ratio greater than that of three to two. Of all the propositions just enumerated Heracleides brought you the proofs. The proposition stated next after these was wrong, viz. that, if a sphere be cut by a plane into unequal parts, the greater segment will have to the less the duplicate ratio of that which the greater surface has to the less. That this is wrong is obvious by what I sent you before; for it included this proposition: If a sphere be cut into unequal parts by a plane at right angles to any diameter in the sphere, the greater segment of the surface will have to the less the same ratio as the greater segment of the diameter has to the less, while the greater segment of the sphere has to the less a ratio less than the duplicate ratio of that which the

---

* τὸ δοθὲν τμᾶμα σφαίρας τῷ δοθέντι τμάματι σφαίρας ὁμοιῶσαι, i.e. to make a segment of a sphere similar to one given segment and equal in content to another given segment. [Cf. *On the Sphere and Cylinder*, II. 5.]

ON SPIRALS. 153

greater surface has to the less, but greater than the sesquialterate* of that ratio. The last of the problems was also wrong, viz. that, if the diameter of any sphere be cut so that the square on the greater segment is triple of the square on the lesser segment, and if through the point thus arrived at a plane be drawn at right angles to the diameter and cutting the sphere, the figure in such a form as is the greater segment of the sphere is the greatest of all the segments which have an equal surface. That this is wrong is also clear from the theorems which I before sent you. For it was there proved that the hemisphere is the greatest of all the segments of a sphere bounded by an equal surface.

After these theorems the following were propounded concerning the cone†. If a section of a right-angled cone [a parabola], in which the diameter [axis] remains fixed, be made to revolve so that the diameter [axis] is the axis [of revolution], let the figure described by the section of the right-angled cone be called a *conoid*. And if a plane touch the conoidal figure and another plane drawn parallel to the tangent plane cut off a segment of the conoid, let the *base* of the segment cut off be defined as the cutting plane, and the *vertex* as the point in which the other plane touches the conoid. Now, if the said figure be cut by a plane at right angles to the axis, it is clear that the section will be a circle; but it needs to be proved that the segment cut off will be half as large again as the cone which has the same base as the segment and equal height. And if two segments be cut off from the conoid by planes drawn in any manner, it is clear that the sections will be sections of acute-angled cones [ellipses] if the cutting planes be not at right angles to the axis; but it needs to be proved that the segments will bear to one another the ratio of the squares on the lines drawn from their vertices parallel to the axis to meet the cutting planes. The proofs of these propositions are not yet sent to you.

After these came the following propositions about the *spiral*,

---

* (λόγον) μείζονα ἢ ἡμιόλιον τοῦ, ὃν ἔχει κ.τ.λ., i.e. a ratio greater than (the ratio of the surfaces)$^{\frac{3}{2}}$. See *On the Sphere and Cylinder*, II. 8.

† This should be presumably 'the *conoid*,' not 'the cone.'

which are as it were another sort of problem having nothing in common with the foregoing; and I have written out the proofs of them for you in this book. They are as follows. If a straight line of which one extremity remains fixed be made to revolve at a uniform rate in a plane until it returns to the position from which it started, and if, at the same time as the straight line revolves, a point move at a uniform rate along the straight line, starting from the fixed extremity, the point will describe a spiral in the plane. I say then that the area bounded by the spiral and the straight line which has returned to the position from which it started is a third part of the circle described with the fixed point as centre and with radius the length traversed by the point along the straight line during the one revolution. And, if a straight line touch the spiral at the extreme end of the spiral, and another straight line be drawn at right angles to the line which has revolved and resumed its position from the fixed extremity of it, so as to meet the tangent, I say that the straight line so drawn to meet it is equal to the circumference of the circle. Again, if the revolving line and the point moving along it make several revolutions and return to the position from which the straight line started, I say that the area added by the spiral in the third revolution will be double of that added in the second, that in the fourth three times, that in the fifth four times, and generally the areas added in the later revolutions will be multiples of that added in the second revolution according to the successive numbers, while the area bounded by the spiral in the first revolution is a sixth part of that added in the second revolution. Also, if on the spiral described in one revolution two points be taken and straight lines be drawn joining them to the fixed extremity of the revolving line, and if two circles be drawn with the fixed point as centre and radii the lines drawn to the fixed extremity of the straight line, and the shorter of the two lines be produced, I say that (1) the area bounded by the circumference of the greater circle in the direction of (the part of) the spiral included between the straight lines, the spiral (itself) and the produced straight line will bear to (2) the area bounded by the circumference of the lesser circle, the same (part of the) spiral and the straight line joining their extremities the ratio which (3) the radius of the lesser circle together with two thirds of the excess of the radius of the greater circle over the radius of the lesser bears to (4) the radius of the lesser circle together with one third of the said excess.

The proofs then of these theorems and others relating to the spiral are given in the present book. Prefixed to them, after the manner usual in other geometrical works, are the propositions necessary to the proofs of them. And here too, as in the books previously published, I assume the following lemma, that, if there be (two) unequal lines or (two) unequal areas, the excess by which the greater exceeds the less can, by being [continually] added to itself, be made to exceed any given magnitude among those which are comparable with [it and with] one another."

Proposition 1.

If a point move at a uniform rate along any line, and two lengths be taken on it, they will be proportional to the times of describing them.

Two unequal lengths are taken on a straight line, and two lengths on another straight line representing the times; and they are proved to be proportional by taking equimultiples of each length and the corresponding time after the manner of Eucl. V. Def. 5.

Proposition 2.

If each of two points on different lines respectively move along them each at a uniform rate, and if lengths be taken, one on each line, forming pairs, such that each pair are described in equal times, the lengths will be proportionals.

This is proved at once by equating the ratio of the lengths taken on one line to that of the times of description, which must also be equal to the ratio of the lengths taken on the other line.

Proposition 3.

Given any number of circles, it is possible to find a straight line greater than the sum of all their circumferences.

For we have only to describe polygons about each and then take a straight line equal to the sum of the perimeters of the polygons.

Proposition 4.

Given two unequal lines, viz. a straight line and the circumference of a circle, it is possible to find a straight line less than the greater of the two lines and greater than the less.

For, by the Lemma, the excess can, by being added a sufficient number of times to itself, be made to exceed the lesser line.

Thus e.g., if $c > l$ (where c is the circumference of the circle and l the length of the straight line), we can find a number n such that
$$n(c - l) > l.$$
Therefore
$$c - l > \frac{l}{n},$$
and
$$c > l + \frac{l}{n} > l.$$

Hence we have only to divide l into n equal parts and add one of them to l. The resulting line will satisfy the condition.

Proposition 5.

Given a circle with centre O, and the tangent to it at a point A, it is possible to draw from O a straight line OPF, meeting the circle in P and the tangent in F, such that, if c be the circumference of any given circle whatever,
$$FP : OP < (\text{arc } AP) : c.$$

Take a straight line, as D, greater than the circumference c. [Prop. 3]

ON SPIRALS.

Through O draw OH parallel to the given tangent, and draw through A a line APH, meeting the circle in P and OH

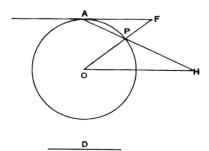

in H, such that the portion PH intercepted between the circle and the line OH may be equal to D^*. Join OP and produce it to meet the tangent in F.

Then $\qquad FP : OP = AP : PH$, by parallels,
$$= AP : D$$
$$< (\text{arc } AP) : c.$$

Proposition 6.

Given a circle with centre O, a chord AB less than the diameter, and OM the perpendicular on AB from O, it is possible to draw a straight line OFP, meeting the chord AB in F and the circle in P, such that
$$FP : PB = D : E,$$
where $D : E$ is any given ratio less than $BM : MO$.

Draw OH parallel to AB, and BT perpendicular to BO meeting OH in T.

Then the triangles BMO, OBT are similar, and therefore
$$BM : MO = OB : BT,$$
whence $\qquad D : E < OB : BT.$

---

* This construction, which is assumed without any explanation as to how it is to be effected, is described in the original Greek thus: "let PH be placed (κείσθω) equal to D, verging (νεύουσα) towards A." This is the usual phraseology used in the type of problem known by the name of νεῦσις.

Suppose that a line PH (greater than BT) is taken such that
$$D : E = OB : PH,$$

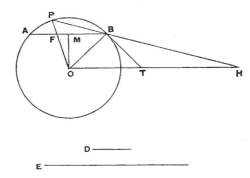

and let PH be so placed that it passes through B and P lies on the circumference of the circle, while H is on the line OH*. (PH will fall outside BT, because $PH > BT$.) Join OP meeting AB in F.

We now have
$$FP : PB = OP : PH$$
$$= OB : PH$$
$$= D : E.$$

Proposition 7.

Given a circle with centre O, a chord AB less than the diameter, and OM the perpendicular on it from O, it is possible to draw from O a straight line OPF, meeting the circle in P and AB produced in F, such that
$$FP : PB = D : E,$$
where $D : E$ is any given ratio greater than $BM : MO$.

Draw OT parallel to AB, and BT perpendicular to BO meeting OT in T.

---

* The Greek phrase is "let PH be placed between the circumference and the straight line (OH) through B." The construction is assumed, like the similar one in the last proposition.

In this case, $D : E > BM : MO$
$> OB : BT$, by similar triangles.

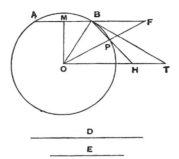

Take a line PH (less than BT) such that
$$D : E = OB : PH,$$
and place PH so that P, H are on the circle and on OT respectively, while HP produced passes through B*.

Then
$$FP : PB = OP : PH$$
$$= D : E.$$

Proposition 8.

Given a circle with centre O, a chord AB less than the diameter, the tangent at B, and the perpendicular OM from O on AB, it is possible to draw from O a straight line OFP, meeting the chord AB in F, the circle in P and the tangent in G, such that
$$FP : BG = D : E,$$
where $D : E$ is any given ratio less than $BM : MO$.

If OT be drawn parallel to AB meeting the tangent at B in T,
$$BM : MO = OB : BT,$$
so that
$$D : E < OB : BT.$$

Take a point C on TB produced such that
$$D : E = OB : BC,$$
whence
$$BC > BT.$$

---

* PH is described in the Greek as νεύουσαν ἐπὶ (*verging to*) the point B. As before the construction is assumed.

Through the points O, T, C describe a circle, and let OB be produced to meet this circle in K.

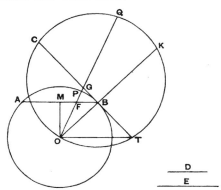

Then, since $BC > BT$, and OB is perpendicular to CT, it is possible to draw from O a straight line OGQ, meeting CT in G and the circle about OTC in Q, such that $GQ = BK$*.

Let OGQ meet AB in F and the original circle in P.

Now $CG \cdot GT = OG \cdot GQ$;

and $OF : OG = BT : GT$,

so that $OF \cdot GT = OG \cdot BT$.

It follows that

$CG \cdot GT : OF \cdot GT = OG \cdot GQ : OG \cdot BT$,

or $CG : OF = GQ : BT$

$= BK : BT$, by construction,

$= BC : OB$

$= BC : OP$.

Hence $OP : OF = BC : CG$,

and therefore $PF : OP = BG : BC$,

or $PF : BG = OP : BC$

$= OB : BC$

$= D : E$.

* The Greek words used are: "it is possible to place another [straight line] GQ equal to KB verging (νεύουσαν) towards O." This particular νεῦσις is discussed by Pappus (p. 298, ed. Hultsch). See the Introduction, chapter v.

Proposition 9.

Given a circle with centre O, a chord AB less than the diameter, the tangent at B, and the perpendicular OM from O on AB, it is possible to draw from O a straight line OPGF, meeting the circle in P, the tangent in G, and AB produced in F, such that

$$FP : BG = D : E,$$

where $D : E$ is any given ratio greater than $BM : MO$.

Let OT be drawn parallel to AB meeting the tangent at B in T.

Then $\qquad D : E > BM : MO$

$\qquad\qquad\qquad > OB : BT$, by similar triangles.

Produce TB to C so that

$$D : E = OB : BC,$$

whence $\qquad BC < BT.$

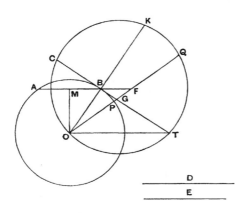

Describe a circle through the points O, T, C, and produce OB to meet this circle in K.

Then, since $TB > BC$, and OB is perpendicular to CT, it is possible to draw from O a line OGQ, meeting CT in G, and the circle about OTC in Q, such that $GQ = BK$*. Let OQ meet the original circle in P and AB produced in F.

We now prove, exactly as in the last proposition, that
$$CG : OF = BK : BT$$
$$= BC : OP.$$

Thus, as before,
$$OP : OF = BC : CG,$$
and
$$OP : PF = BC : BG,$$
whence
$$PF : BG = OP : BC$$
$$= OB : BC$$
$$= D : E.$$

Proposition 10.

If $A_1, A_2, A_3, \ldots A_n$ be n lines forming an ascending arithmetical progression in which the common difference is equal to A_1, the least term, then

$$(n+1) A_n^2 + A_1 (A_1 + A_2 + \ldots + A_n) = 3 (A_1^2 + A_2^2 + \ldots + A_n^2).$$

[Archimedes' proof of this proposition is given above, p. 107–9, and it is there pointed out that the result is equivalent to

$$1^2 + 2^2 + 3^2 + \ldots + n^2 = \frac{n(n+1)(2n+1)}{6}.]$$

COR. 1. *It follows from this proposition that*
$$n \cdot A_n^2 < 3 (A_1^2 + A_2^2 + \ldots + A_n^2),$$
and also that
$$n \cdot A_n^2 > 3 (A_1^2 + A_2^2 + \ldots + A_{n-1}^2).$$

[For the proof of the latter inequality see p. 109 above.]

COR. 2. *All the results will equally hold if similar figures are substituted for squares.*

---

* See the note on the last proposition.

Proposition 11.

If $A_1, A_2, \ldots A_n$ be n lines forming an ascending arithmetical progression [in which the common difference is equal to the least term A_1]*, then

$$(n-1)A_n^2 : (A_n^2 + A_{n-1}^2 + \ldots + A_2^2)$$
$$< A_n^2 : \{A_n \cdot A_1 + \tfrac{1}{3}(A_n - A_1)^2\};$$

but

$$(n-1)A_n^2 : (A_{n-1}^2 + A_{n-2}^2 + \ldots + A_1^2)$$
$$> A_n^2 : \{A_n \cdot A_1 + \tfrac{1}{3}(A_n - A_1)^2\}.$$

[Archimedes sets out the terms side by side in the manner shown in the figure, where $BC = A_n$, $DE = A_{n-1}, \ldots RS = A_1$, and produces DE, $FG, \ldots RS$ until they are respectively equal to BC or A_n, so that EH, $GI, \ldots SU$ in the figure are respectively equal to $A_1, A_2 \ldots A_{n-1}$. He further measures lengths BK, DL, $FM, \ldots PV$ along BC, DE, $FG, \ldots PQ$ respectively each equal to RS.

The figure makes the relations between the terms easier to see with the eye, but the use of so large a number of letters makes the proof somewhat difficult to follow, and it may be more clearly represented as follows.]

It is evident that $(A_n - A_1) = A_{n-1}$.

The following proportion is therefore obviously true, viz.

$$(n-1)A_n^2 : (n-1)(A_n \cdot A_1 + \tfrac{1}{3} A_{n-1}^2)$$
$$= A_n^2 : \{A_n \cdot A_1 + \tfrac{1}{3}(A_n - A_1)^2\}.$$

---

* The proposition is true even when the common difference is not equal to A_1, and is assumed in the more general form in Props. 25 and 26. But, as Archimedes' proof assumes the equality of A_1 and the common difference, the words are here inserted to prevent misapprehension.

In order therefore to prove the desired result, we have only to show that

$$(n-1)A_n \cdot A_1 + \tfrac{1}{3}(n-1)A_{n-1}^2 < (A_n^2 + A_{n-1}^2 + \ldots + A_2^2)$$
but
$$> (A_{n-1}^2 + A_{n-2}^2 + \ldots + A_1^2).$$

I. To prove the first inequality, we have
$$(n-1)A_n \cdot A_1 + \tfrac{1}{3}(n-1)A_{n-1}^2$$
$$= (n-1)A_1^2 + (n-1)A_1 \cdot A_{n-1} + \tfrac{1}{3}(n-1)A_{n-1}^2 \ldots (1).$$

And
$$A_n^2 + A_{n-1}^2 + \ldots + A_2^2$$
$$= (A_{n-1} + A_1)^2 + (A_{n-2} + A_1)^2 + \ldots + (A_1 + A_1)^2$$
$$= (A_{n-1}^2 + A_{n-2}^2 + \ldots + A_1^2)$$
$$\quad + (n-1)A_1^2$$
$$\quad + 2A_1(A_{n-1} + A_{n-2} + \ldots + A_1)$$
$$= (A_{n-1}^2 + A_{n-2}^2 + \ldots + A_1^2)$$
$$\quad + (n-1)A_1^2$$
$$\quad + A_1\{A_{n-1} + A_{n-2} + A_{n-3} + \ldots + A_1$$
$$\qquad\qquad + A_1 \;\; + A_2 \;\; + \ldots + A_{n-2} + A_{n-1}\}$$
$$= (A_{n-1}^2 + A_{n-2}^2 + \ldots + A_1^2)$$
$$\quad + (n-1)A_1^2$$
$$\quad + nA_1 \cdot A_{n-1} \ldots\ldots\ldots\ldots\ldots\ldots\ldots\ldots\ldots\ldots\ldots\ldots (2).$$

Comparing the right-hand sides of (1) and (2), we see that $(n-1)A_1^2$ is common to both sides, and
$$(n-1)A_1 \cdot A_{n-1} < nA_1 \cdot A_{n-1},$$
while, by Prop. 10, Cor. 1,
$$\tfrac{1}{3}(n-1)A_{n-1}^2 < A_{n-1}^2 + A_{n-2}^2 + \ldots + A_1^2.$$
It follows therefore that
$$(n-1)A_n \cdot A_1 + \tfrac{1}{3}(n-1)A_{n-1}^2 < (A_n^2 + A_{n-1}^2 + \ldots + A_2^2);$$
and hence the first part of the proposition is proved.

II. We have now, in order to prove the second result, to show that
$$(n-1)A_n \cdot A_1 + \tfrac{1}{3}(n-1)A_{n-1}^2 > (A_{n-1}^2 + A_{n-2}^2 + \ldots + A_1^2).$$

The right-hand side is equal to

$$(A_{n-2}+A_1)^2+(A_{n-3}+A_1)^2+\ldots+(A_1+A_1)^2+A_1^2$$
$$=A_{n-2}^2+A_{n-3}^2+\ldots+A_1^2$$
$$+(n-1)A_1^2$$
$$+2A_1(A_{n-2}+A_{n-3}+\ldots+A_1)$$
$$=(A_{n-2}^2+A_{n-3}^2+\ldots+A_1^2)$$
$$+(n-1)A_1^2$$
$$+A_1\left\{\begin{matrix}A_{n-2}+A_{n-3}+\ldots+A_1\\+A_1\phantom{_{n-2}}+A_2\phantom{_{n-3}}+\ldots+A_{n-2}\end{matrix}\right\}$$
$$=(A_{n-2}^2+A_{n-3}^2+\ldots+A_1^2)$$
$$+(n-1)A_1^2$$
$$+(n-2)A_1.A_{n-1}\ldots\ldots\ldots\ldots\ldots\ldots\ldots\ldots(3).$$

Comparing this expression with the right-hand side of (1) above, we see that $(n-1)A_1^2$ is common to both sides, and

$$(n-1)A_1.A_{n-1} > (n-2)A_1.A_{n-1},$$

while, by Prop. 10, Cor. 1,

$$\tfrac{1}{3}(n-1)A_{n-1}^2 > (A_{n-2}^2+A_{n-3}^2+\ldots+A_1^2).$$

Hence

$$(n-1)A_n.A_1+\tfrac{1}{3}(n-1)A_{n-1}^2 > (A_{n-1}^2+A_{n-2}^2+\ldots+A_1^2);$$

and the second required result follows.

COR. *The results in the above proposition are equally true if similar figures be substituted for squares on the several lines.*

DEFINITIONS.

1. If a straight line drawn in a plane revolve at a uniform rate about one extremity which remains fixed and return to the position from which it started, and if, at the same time as the line revolves, a point move at a uniform rate along the straight line beginning from the extremity which remains fixed, the point will describe a **spiral** (ἕλιξ) in the plane.

2. Let the extremity of the straight line which remains

fixed while the straight line revolves be called the **origin*** (ἀρχά) of the spiral.

3. And let the position of the line from which the straight line began to revolve be called the **initial line*** in the revolution (ἀρχὰ τᾶς περιφορᾶς).

4. Let the length which the point that moves along the straight line describes in one revolution be called the **first distance**, that which the same point describes in the second revolution the **second distance**, and similarly let the distances described in further revolutions be called after the number of the particular revolution.

5. Let the area bounded by the spiral described in the first revolution and the *first distance* be called the **first area**, that bounded by the spiral described in the second revolution and the *second distance* the **second area**, and similarly for the rest in order.

6. If from the origin of the spiral any straight line be drawn, let that side of it which is in the same direction as that of the revolution be called **forward** (προαγούμενα), and that which is in the other direction **backward** (ἑπόμενα).

7. Let the circle drawn with the *origin* as centre and the *first distance* as radius be called the **first circle**, that drawn with the same centre and twice the radius the **second circle**, and similarly for the succeeding circles.

Proposition 12.

If any number of straight lines drawn from the origin to meet the spiral make equal angles with one another, the lines will be in arithmetical progression.

[The proof is obvious.]

---

* The literal translation would of course be the "beginning of the spiral" and "the beginning of the revolution" respectively. But the modern names will be more suitable for use later on, and are therefore employed here.

Proposition 13.

If a straight line touch the spiral, it will touch it in one point only.

Let O be the origin of the spiral, and BC a tangent to it.

If possible, let BC touch the spiral in two points P, Q. Join OP, OQ, and bisect the angle POQ by the straight line OR meeting the spiral in R.

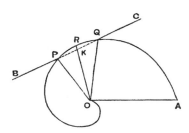

Then [Prop. 12] OR is an arithmetic mean between OP and OQ, or
$$OP + OQ = 2OR.$$
But in any triangle POQ, if the bisector of the angle POQ meets PQ in K,
$$OP + OQ > 2OK^*.$$
Therefore $OK < OR$, and it follows that some point on BC between P and Q lies within the spiral. Hence BC cuts the spiral; which is contrary to the hypothesis.

Proposition 14.

If O be the origin, and P, Q two points on the first turn of the spiral, and if OP, OQ produced meet the 'first circle' $AKP'Q'$ in P', Q' respectively, OA being the initial line, then
$$OP : OQ = (\text{arc } AKP') : (\text{arc } AKQ').$$

For, while the revolving line OA moves about O, the point A on it moves uniformly along the circumference of the circle

---

* This is assumed as a known proposition; but it is easily proved.

$AKP'Q'$, and at the same time the point describing the spiral moves uniformly along OA.

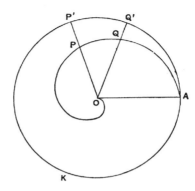

Thus, while A describes the arc AKP', the moving point on OA describes the length OP, and, while A describes the arc AKQ', the moving point on OA describes the distance OQ.

Hence $OP : OQ = (\text{arc } AKP') : (\text{arc } AKQ')$.   [Prop. 2]

Proposition 15.

If P, Q be points on the second turn of the spiral, and OP, OQ meet the 'first circle' $AKP'Q'$ in P', Q', as in the last proposition, and if c be the circumference of the first circle, then

$$OP : OQ = c + (\text{arc } AKP') : c + (\text{arc } AKQ').$$

For, while the moving point on OA describes the distance OP, the point A describes the whole of the circumference of the 'first circle' together with the arc AKP'; and, while the moving point on OA describes the distance OQ, the point A describes the whole circumference of the 'first circle' together with the arc AKQ'.

Cor. Similarly, if P, Q are on the nth turn of the spiral,

$$OP : OQ = (n-1)c + (\text{arc } AKP') : (n-1)c + (\text{arc } AKQ').$$

Propositions 16, 17.

If BC be the tangent at P, any point on the spiral, PC being the 'forward' part of BC, and if OP be joined, the angle OPC is obtuse while the angle OPB is acute.

I. Suppose *P* to be on the first turn of the spiral.

Let *OA* be the initial line, *AKP'* the 'first circle.' Draw the circle *DLP* with centre *O* and radius *OP*, meeting *OA* in *D*. This circle must then, in the 'forward' direction from *P*,

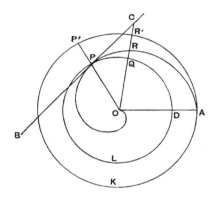

fall within the spiral, and in the 'backward' direction outside it, since the radii vectores of the spiral are on the 'forward' side greater, and on the 'backward' side less, than *OP*. Hence the angle *OPC* cannot be acute, since it cannot be less than the angle between *OP* and the tangent to the circle at *P*, which is a right angle.

It only remains therefore to prove that *OPC* is not a right angle.

If possible, let it be a right angle. *BC* will then touch the circle at *P*.

Therefore [Prop. 5] it is possible to draw a line *OQC* meeting the circle through *P* in *Q* and *BC* in *C*, such that
$$CQ : OQ < (\text{arc } PQ) : (\text{arc } DLP) \ldots\ldots\ldots\ldots(1).$$

Suppose that OC meets the spiral in R and the 'first circle' in R'; and produce OP to meet the 'first circle' in P'.

From (1) it follows, *componendo*, that

$$CO : OQ < (\text{arc } DLQ) : (\text{arc } DLP)$$
$$< (\text{arc } AKR') : (\text{arc } AKP')$$
$$< OR : OP. \qquad \text{[Prop. 14]}$$

But this is impossible, because $OQ = OP$, and $OR < OC$.

Hence the angle OPC is not a right angle. It was also proved not to be acute.

Therefore the angle OPC is obtuse, and the angle OPB consequently acute.

II. If P is on the second, or the nth turn, the proof is the same, except that in the proportion (1) above we have to substitute for the arc DLP an arc equal to $(p + \text{arc } DLP)$ or $(\overline{n-1} \cdot p + \text{arc } DLP)$, where p is the perimeter of the circle

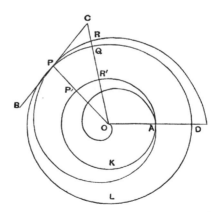

DLP through P. Similarly, in the later steps, p or $(n-1)p$ will be added to each of the arcs DLQ and DLP, and c or $(n-1)c$ to each of the arcs AKR', AKP', where c is the circumference of the 'first circle' AKP'.

Propositions 18, 19.

I. *If OA be the initial line, A the end of the first turn of the spiral, and if the tangent to the spiral at A be drawn, the straight line OB drawn from O perpendicular to OA will meet the said tangent in some point B, and OB will be equal to the circumference of the 'first circle.'*

II. *If A' be the end of the second turn, the perpendicular OB will meet the tangent at A' in some point B', and OB' will be equal to 2 (circumference of 'second circle').*

III. *Generally, if A_n be the end of the nth turn, and OB meet the tangent at A_n in B_n, then*

$$OB_n = nc_n,$$

where c_n is the circumference of the 'nth circle.'

I. Let AKC be the 'first circle.' Then, since the 'backward' angle between OA and the tangent at A is acute [Prop. 16], the tangent will meet the 'first circle' in a second point C. And the angles CAO, BOA are together less than two right angles; therefore OB will meet AC produced in some point B.

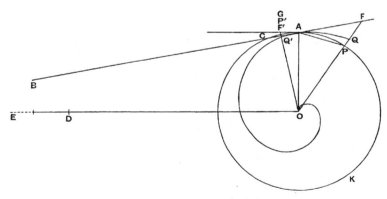

Then, if c be the circumference of the first circle, we have to prove that
$$OB = c.$$

If not, OB must be either greater or less than c.

(1) If possible, suppose $OB > c$.

Measure along OB a length OD less than OB but greater than c.

We have then a circle AKC, a chord AC in it less than the diameter, and a ratio $AO:OD$ which is greater than the ratio $AO:OB$ or (what is, by similar triangles, equal to it) the ratio of $\frac{1}{2}AC$ to the perpendicular from O on AC. Therefore [Prop. 7] we can draw a straight line OPF, meeting the circle in P and CA produced in F, such that
$$FP:PA = AO:OD.$$
Thus, alternately, since $AO = PO$,
$$FP:PO = PA:OD$$
$$< (\text{arc } PA):c,$$
since (arc PA) > PA, and $OD > c$.

Componendo,
$$FO:PO < (c + \text{arc } PA):c$$
$$< OQ:OA,$$
where OF meets the spiral in Q. [Prop. 15]

Therefore, since $OA = OP$, $FO < OQ$; which is impossible.

Hence $OB \not> c$.

(2) If possible, suppose $OB < c$.

Measure OE along OB so that OE is greater than OB but less than c.

In this case, since the ratio $AO:OE$ is less than the ratio $AO:OB$ (or the ratio of $\frac{1}{2}AC$ to the perpendicular from O on AC), we can [Prop. 8] draw a line $OF'P'G$, meeting AC in F', the circle in P', and the tangent at A to the circle in G, such that
$$F'P':AG = AO:OE.$$
Let $OP'G$ cut the spiral in Q'.

Then we have, alternately,
$$F'P':P'O = AG:OE$$
$$> (\text{arc } AP'):c,$$
because $AG > (\text{arc } AP')$, and $OE < c$.

Therefore
$$F'O : P'O < (\text{arc } AKP') : c$$
$$< OQ' : OA. \quad\quad \text{[Prop. 14]}$$
But this is impossible, since $OA = OP'$, and $OQ' < OF'$.

Hence $OB \not< c$.

Since therefore OB is neither greater nor less than c,
$$OB = c.$$

II. Let $A'K'C'$ be the 'second circle,' $A'C'$ being the tangent to the spiral at A' (which will cut the second circle, since the 'backward' angle $OA'C'$ is acute). Thus, as before, the perpendicular OB' to OA' will meet $A'C'$ produced in some point B'.

If then c' is the circumference of the 'second circle,' we have to prove that $OB' = 2c'$.

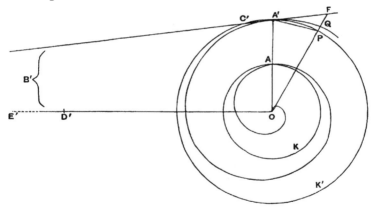

For, if not, OB' must be either greater or less than $2c'$.

(1) If possible, suppose $OB' > 2c'$.

Measure OD' along OB' so that OD' is less than OB' but greater than $2c'$.

Then, as in the case of the 'first circle' above, we can draw a straight line OPF meeting the 'second circle' in P and $C'A'$ produced in F, such that
$$FP : PA' = A'O : OD'.$$

Let OF meet the spiral in Q.

We now have, since $A'O = PO$,

$$FP : PO = PA' : OD'$$
$$< (\text{arc } A'P) : 2c',$$

because (arc $A'P$) $> A'P$ and $OD' > 2c'$.

Therefore $FO : PO < (2c' + \text{arc } A'P) : 2c'$
$< OQ : OA'.$ [Prop. 15, Cor.]

Hence $FO < OQ$; which is impossible.

Thus $OB' \not> 2c'$.

Similarly, as in the case of the 'first circle', we can prove that

$$OB' \not< 2c'.$$

Therefore $OB' = 2c'$.

III. Proceeding, in like manner, to the 'third' and succeeding circles, we shall prove that

$$OB_n = nc_n.$$

Proposition 20.

I. *If P be any point on the first turn of the spiral and OT be drawn perpendicular to OP, OT will meet the tangent at P to the spiral in some point T; and, if the circle drawn with centre O and radius OP meet the initial line in K, then OT is equal to the arc of this circle between K and P measured in the 'forward' direction of the spiral.*

II. *Generally, if P be a point on the nth turn, and the notation be as before, while p represents the circumference of the circle with radius OP,*

$OT = (n - 1) p + \text{arc } KP$ *(measured 'forward')*.

I. Let P be a point on the first turn of the spiral, OA the initial line, PR the tangent at P taken in the 'backward' direction.

Then [Prop. 16] the angle OPR is acute. Therefore PR

meets the circle through P in some point R; and also OT will meet PR produced in some point T.

If now OT is not equal to the arc KRP, it must be either greater or less.

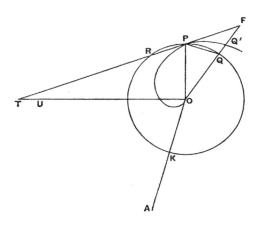

(1) If possible, let OT be greater than the arc KRP.

Measure OU along OT less than OT but greater than the arc KRP.

Then, since the ratio $PO : OU$ is greater than the ratio $PO : OT$, or (what is, by similar triangles, equal to it) the ratio of $\frac{1}{2}PR$ to the perpendicular from O on PR, we can draw a line OQF, meeting the circle in Q and RP produced in F, such that

$$FQ : PQ = PO : OU. \qquad \text{[Prop. 7]}$$

Let OF meet the spiral in Q'.

We have then
$$FQ : QO = PQ : OU$$
$$< (\text{arc } PQ) : (\text{arc } KRP), \text{ by hypothesis.}$$

Componendo,
$$FO : QO < (\text{arc } KRQ) : (\text{arc } KRP)$$
$$< OQ' : OP. \qquad \text{[Prop. 14]}$$

But $QO = OP$.

Therefore $FO < OQ'$; which is impossible.

Hence $OT \not> (\text{arc } KRP)$.

(2) The proof that $OT \not< (\text{arc } KRP)$ follows the method of Prop. 18, I. (2), exactly as the above follows that of Prop. 18, I. (1).

Since then OT is neither greater nor less than the arc KRP, it is equal to it.

II. If P be on the second turn, the same method shows that
$$OT = p + (\text{arc } KRP);$$
and, similarly, we have, for a point P on the nth turn,
$$OT = (n-1)p + (\text{arc } KRP).$$

Propositions 21, 22, 23.

Given an area bounded by any arc of a spiral and the lines joining the extremities of the arc to the origin, it is possible to circumscribe about the area one figure, and to inscribe in it another figure, each consisting of similar sectors of circles, and such that the circumscribed figure exceeds the inscribed by less than any assigned area.

For let BC be any arc of the spiral, O the origin. Draw the circle with centre O and radius OC, where C is the 'forward' end of the arc.

Then, by bisecting the angle BOC, bisecting the resulting angles, and so on continually, we shall ultimately arrive at an angle COr cutting off a sector of the circle less than any assigned area. Let COr be this sector.

Let the other lines dividing the angle BOC into equal parts meet the spiral in P, Q, and let Or meet it in R. With O as centre and radii OB, OP, OQ, OR respectively describe arcs of circles Bp', bBq', pQr', qRc', each meeting the adjacent radii as shown in the figure. In each case the arc in the 'forward' direction from each point will fall within, and the arc in the 'backward' direction outside, the spiral.

We have now a circumscribed figure and an inscribed figure each consisting of similar sectors of circles. To compare their areas, we take the successive sectors of each, beginning from OC, and compare them.

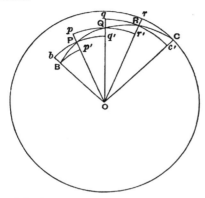

The sector OCr in the circumscribed figure stands alone.

And
$$(\text{sector } ORq) = (\text{sector } ORc'),$$
$$(\text{sector } OQp) = (\text{sector } OQr'),$$
$$(\text{sector } OPb) = (\text{sector } OPq'),$$

while the sector OBp' in the inscribed figure stands alone.

Hence, if the equal sectors be taken away, the difference between the circumscribed and inscribed figures is equal to the difference between the sectors OCr and OBp'; and this difference is less than the sector OCr, which is itself less than any assigned area.

The proof is exactly the same whatever be the number of angles into which the angle BOC is divided, the only difference being that, when the arc begins from the origin, the smallest sectors OPb, OPq' in each figure are equal, and there is therefore no inscribed sector standing by itself, so that the difference between the circumscribed and inscribed figures is equal to the sector OCr itself.

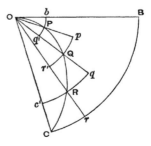

Thus the proposition is universally true.

COR. Since the area bounded by the spiral is intermediate in magnitude between the circumscribed and inscribed figures, it follows that

(1) *a figure can be circumscribed to the area such that it exceeds the area by less than any assigned space,*

(2) *a figure can be inscribed such that the area exceeds it by less than any assigned space.*

Proposition 24.

The area bounded by the first turn of the spiral and the initial line is equal to one-third of the 'first circle' $[= \frac{1}{3}\pi(2\pi a)^2$, where the spiral is $r = a\theta$].

[The same proof shows equally that, *if OP be any radius vector in the first turn of the spiral, the area of the portion of the spiral bounded thereby is equal to one-third of that sector of the circle drawn with radius OP which is bounded by the initial line and OP, measured in the 'forward' direction from the initial line.*]

Let O be the origin, OA the initial line, A the extremity of the first turn.

Draw the 'first circle,' i.e. the circle with O as centre and OA as radius.

Then, if C_1 be the area of the first circle, R_1 that of the first turn of the spiral bounded by OA, we have to prove that

$$R_1 = \tfrac{1}{3}C_1.$$

For, if not, R_1 must be either greater or less than C_1.

I. If possible, suppose $R_1 < \tfrac{1}{3}C_1$.

We can then circumscribe a figure about R_1 made up of similar sectors of circles such that, if F be the area of this figure,

$$F - R_1 < \tfrac{1}{3}C_1 - R_1,$$

whence $F < \tfrac{1}{3}C_1$.

Let $OP, OQ, \ldots$ be the radii of the circular sectors, beginning from the smallest. The radius of the largest is of course OA.

The radii then form an ascending arithmetical progression in which the common difference is equal to the least term OP. If n be the number of the sectors, we have [by Prop. 10, Cor. 1]
$$n \cdot OA^2 < 3(OP^2 + OQ^2 + \ldots + OA^2);$$

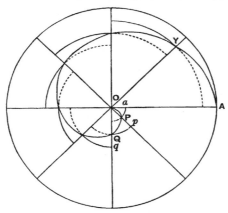

and, since the similar sectors are proportional to the squares on their radii, it follows that
$$C_1 < 3F,$$
or $\qquad\qquad F > \tfrac{1}{3}C_1.$

But this is impossible, since F was less than $\tfrac{1}{3}C_1$.

Therefore $\qquad R_1 \not> \tfrac{1}{3}C_1.$

II. If possible, suppose $R_1 > \tfrac{1}{3}C_1$.

We can then *inscribe* a figure made up of similar sectors of circles such that, if f be its area,
$$R_1 - f < R_1 - \tfrac{1}{3}C_1,$$
whence $f > \tfrac{1}{3}C_1$.

If there are $(n-1)$ sectors, their radii, as $OP, OQ, \ldots$, form an ascending arithmetical progression in which the least term is equal to the common difference, and the greatest term, as OY, is equal to $(n-1)\,OP$.

Thus [Prop. 10, Cor. 1]
$$n \cdot OA^2 > 3(OP^2 + OQ^2 + \ldots + OY^2),$$
whence $\qquad C_1 > 3f,$
or $\qquad f < \tfrac{1}{3}C_1;$
which is impossible, since $f > \tfrac{1}{3}C_1$.

Therefore $\qquad R_1 \not> \tfrac{1}{3}C_1.$

Since then R_1 is neither greater nor less than $\tfrac{1}{3}C_1$,
$$R_1 = \tfrac{1}{3}C_1.$$

[Archimedes does not actually find the area of the spiral cut off by the radius vector OP, where P is any point on the first turn; but, in order to do this, we have only to substitute

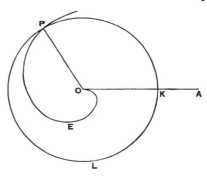

in the above proof the area of the sector KLP of the circle drawn with O as centre and OP as radius for the area C_1 of the 'first circle', while the two figures made up of similar sectors have to be circumscribed about and inscribed in the portion OEP of the spiral. The same method of proof then applies exactly, and the area of OEP is seen to be $\tfrac{1}{3}$ (sector KLP).

We can prove also, by the same method, that, if P be a point on the second, or any later turn, as the nth, the *complete area described by the radius vector* from the beginning up to the time when it reaches the position OP is, if C denote the area of the complete circle with O as centre and OP as radius, $\tfrac{1}{3}(C + \text{sector } KLP)$ or $\tfrac{1}{3}(\overline{n-1} \cdot C + \text{sector } KLP)$ respectively.

The area so described by the radius vector is of course not the same thing as the area bounded by the last complete turn

ON SPIRALS. 181

of the spiral ending at P and the intercepted portion of the radius vector OP. Thus, suppose R_1 to be the area bounded by the first turn of the spiral and OA_1 (the first turn ending at A_1 on the initial line), R_2 the area *added* to this by the second complete turn ending at A_2 on the initial line, and so on. R_1 has then been described *twice* by the radius vector when it arrives at the position OA_2; when the radius vector arrives at the position OA_3, it has described R_1 three times, the ring R_2 twice, and the ring R_3 once; and so on.

Thus, generally, if C_n denote the area of the 'nth circle,' we shall have

$$\tfrac{1}{3}nC_n = R_n + 2R_{n-1} + 3R_{n-2} + \ldots + nR_1,$$

while the actual area bounded by the outside, or the complete nth, turn and the intercepted portion of OA_n will be equal to

$$R_n + R_{n-1} + R_{n-2} + \ldots + R_1.$$

It can now be seen that the results of the later Props. 25 and 26 may be obtained from the extension of Prop. 24 just given.

To obtain the general result of Prop. 26, suppose BC to be an arc on any turn whatever of the spiral, being itself less than a complete turn, and suppose B to be beyond A_n the extremity of the nth complete turn, while C is 'forward' from B.

Let $\dfrac{p}{q}$ be the fraction of a turn between the end of the nth turn and the point B.

Then the *area described by the radius vector* up to the position OB (starting from the beginning of the spiral) is equal to

$$\tfrac{1}{3}\left(n + \tfrac{p}{q}\right)(\text{circle with rad. } OB).$$

Also the *area described by the radius vector* from the beginning up to the position OC is

$$\tfrac{1}{3}\left\{\left(n + \tfrac{p}{q}\right)(\text{circle with rad. } OC) + (\text{sector } B'MC)\right\}.$$

The area bounded by OB, OC and the portion BEC of the spiral is equal to the difference between these two expressions; and, since the circles are to one another as OB^2 to OC^2, the difference may be expressed as

$$\tfrac{1}{3}\left\{\left(n+\frac{p}{q}\right)\left(1-\frac{OB^2}{OC^2}\right)(\text{circle with rad. } OC) + (\text{sector } B'MC)\right\}.$$

But, by Prop. 15, Cor.,

$$\left(n+\frac{p}{q}\right)(\text{circle } B'MC) : \left\{\left(n+\frac{p}{q}\right)(\text{circle } B'MC) + (\text{sector } B'MC)\right\}$$
$$= OB : OC,$$

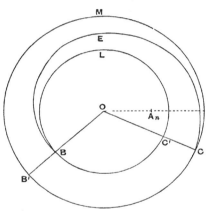

so that

$$\left(n+\frac{p}{q}\right)(\text{circle } B'MC) : (\text{sector } B'MC) = OB : (OC - OB).$$

Thus $\dfrac{\text{area } BEC}{\text{sector } B'MC} = \tfrac{1}{3}\left\{\left(\dfrac{OB}{OC-OB}\right)\left(1-\dfrac{OB^2}{OC^2}\right)+1\right\}$

$$= \tfrac{1}{3} \cdot \frac{OB(OC+OB)+OC^2}{OC^2}$$

$$= \frac{OC \cdot OB + \tfrac{1}{3}(OC-OB)^2}{OC^2}.$$

The result of Prop. 25 is a particular case of this, and the result of Prop. 27 follows immediately, as shown under that proposition.]

Propositions 25, 26, 27.

[Prop. 25.] *If A_2 be the end of the second turn of the spiral, the area bounded by the second turn and OA_2 is to the area of the 'second circle' in the ratio of 7 to 12, being the ratio of $\{r_2 r_1 + \frac{1}{3}(r_2 - r_1)^2\}$ to r_2^2, where r_1, r_2 are the radii of the 'first' and 'second' circles respectively.*

[Prop. 26.] *If BC be any arc measured in the 'forward' direction on any turn of a spiral, not being greater than the complete turn, and if a circle be drawn with O as centre and OC as radius meeting OB in B', then*

(area of spiral between OB, OC) : (sector $OB'C$)
$$= \{OC \cdot OB + \tfrac{1}{3}(OC - OB)^2\} : OC^2.$$

[Prop. 27.] *If R_1 be the area of the first turn of the spiral bounded by the initial line, R_2 the area of the ring added by the second complete turn, R_3 that of the ring added by the third turn, and so on, then*

$$R_3 = 2R_2,\ R_4 = 3R_2,\ R_5 = 4R_2, \ldots, R_n = (n-1)R_2.$$

Also $\qquad\qquad R_2 = 6R_1.$

[Archimedes' proof of Prop. 25 is, *mutatis mutandis*, the same as his proof of the more general Prop. 26. The latter will accordingly be given here, and applied to Prop. 25 as a particular case.]

Let BC be an arc measured in the 'forward' direction on any turn of the spiral, CKB' the circle drawn with O as centre and OC as radius.

Take a circle such that the square of its radius is equal to $OC \cdot OB + \frac{1}{3}(OC - OB)^2$, and let σ be a sector in it whose central angle is equal to the angle BOC.

Thus $\quad \sigma : $ (sector $OB'C$) $= \{OC \cdot OB + \frac{1}{3}(OC - OB)^2\} : OC^2,$

and we have therefore to prove that

(area of spiral OBC) $= \sigma$.

For, if not, the area of the spiral OBC (which we will call S) must be either greater or less than σ.

I. Suppose, if possible, $S < \sigma$.

Circumscribe to the area S a figure made up of similar sectors of circles, such that, if F be the area of the figure,
$$F - S < \sigma - S,$$
whence $F < \sigma.$

Let the radii of the successive sectors, starting from OB, be $OP, OQ, \ldots OC$. Produce $OP, OQ, \ldots$ to meet the circle $CKB', \ldots$

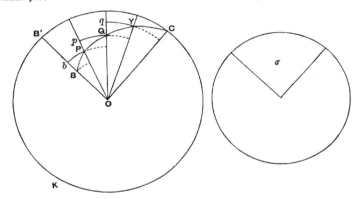

If then the lines $OB, OP, OQ, \ldots OC$ be n in number, the number of sectors in the circumscribed figure will be $(n-1)$, and the sector $OB'C$ will also be divided into $(n-1)$ equal sectors. Also $OB, OP, OQ, \ldots OC$ will form an ascending arithmetical progression of n terms.

Therefore [see Prop. 11 and Cor.]
$$(n-1)OC^2 : (OP^2 + OQ^2 + \ldots + OC^2)$$
$$< OC^2 : \{OC \cdot OB + \tfrac{1}{3}(OC - OB)^2\}$$
$$< (\text{sector } OB'C) : \sigma, \text{ by hypothesis.}$$

Hence, since similar sectors are as the squares of their radii,
$$(\text{sector } OB'C) : F < (\text{sector } OB'C) : \sigma,$$
so that $F > \sigma.$

But this is impossible, because $F < \sigma$.

Therefore $S \not< \sigma.$

II. Suppose, if possible, $S > \sigma$.

Inscribe in the area S a figure made up of similar sectors of circles such that, if f be its area,
$$S - f < S - \sigma,$$
whence $\qquad f > \sigma.$

Suppose $OB, OP, \ldots OY$ to be the radii of the successive sectors making up the figure f, being $(n-1)$ in number.

We shall have in this case [see Prop. 11 and Cor.]
$$(n-1) OC^2 : (OB^2 + OP^2 + \ldots + OY^2)$$
$$> OC^2 : \{OC \cdot OB + \tfrac{1}{3}(OC - OB)^2\},$$
whence $\quad$ (sector $OB'C$) $: f >$ (sector $OB'C$) $: \sigma$,
so that $\qquad f < \sigma.$
But this is impossible, because $f > \sigma$.

Therefore $\qquad S \not> \sigma.$

Since then S is neither greater nor less than σ, it follows that
$$S = \sigma.$$

In the particular case where B coincides with A_1, the end of the first turn of the spiral, and C with A_2, the end of the second turn, the sector $OB'C$ becomes the complete 'second circle,' that, namely, with OA_2 (or r_2) as radius.

Thus

(area of spiral bounded by OA_2) : ('second circle')
$$= \{r_2 r_1 + \tfrac{1}{3}(r_2 - r_1)^2\} : r_2^2$$
$$= (2 + \tfrac{1}{3}) : 4 \quad \text{(since } r_2 = 2r_1\text{)}$$
$$= 7 : 12.$$

Again, the area of the spiral bounded by OA_2 is equal to $R_1 + R_2$ (i.e. the area bounded by the first turn and OA_1, together with the ring added by the second turn). Also the 'second circle' is four times the 'first circle,' and therefore equal to $12 R_1$.

Hence $\qquad (R_1 + R_2) : 12 R_1 = 7 : 12,$
or $\qquad R_1 + R_2 = 7 R_1.$
Thus $\qquad R_2 = 6 R_1 \ldots\ldots\ldots\ldots\ldots\ldots(1).$

Next, for the third turn, we have

$$(R_1 + R_2 + R_3) : (\text{'third circle'}) = \{r_3 r_2 + \tfrac{1}{3}(r_3 - r_2)^2\} : r_3^2$$
$$= (3 \cdot 2 + \tfrac{1}{3}) : 3^2$$
$$= 19 : 27,$$

and $(\text{'third circle'}) = 9 \,(\text{'first circle'})$
$$= 27 R_1;$$

therefore $\quad R_1 + R_2 + R_3 = 19 R_1,$

and, by (1) above, it follows that

$$R_3 = 12 R_1$$
$$= 2 R_2 \quad \dots\dots\dots\dots\dots\dots\dots(2),$$

and so on.

Generally, we have

$$(R_1 + R_2 + \ldots + R_n) : (n\text{th circle}) = \{r_n r_{n-1} + \tfrac{1}{3}(r_n - r_{n-1})^2\} : r_n^2,$$
$$(R_1 + R_2 + \ldots + R_{n-1}) : (\overline{n-1}\text{th circle})$$
$$= \{r_{n-1} r_{n-2} + \tfrac{1}{3}(r_{n-1} - r_{n-2})^2\} : r_{n-1}^2,$$

and $\quad (n\text{th circle}) : (\overline{n-1}\text{th circle}) = r_n^2 : r_{n-1}^2.$

Therefore

$$(R_1 + R_2 + \ldots + R_n) : (R_1 + R_2 + \ldots + R_{n-1})$$
$$= \{n(n-1) + \tfrac{1}{3}\} : \{(n-1)(n-2) + \tfrac{1}{3}\}$$
$$= \{3n(n-1) + 1\} : \{3(n-1)(n-2) + 1\}.$$

Dirimendo,

$R_n : (R_1 + R_2 + \ldots + R_{n-1})$
$$= 6(n-1) : \{3(n-1)(n-2) + 1\} \quad \dots\dots\dots(\alpha).$$

Similarly

$R_{n-1} : (R_1 + R_2 + \ldots + R_{n-2}) = 6(n-2) : \{3(n-2)(n-3) + 1\},$

from which we derive

$$R_{n-1} : (R_1 + R_2 + \ldots + R_{n-1})$$
$$= 6(n-2) : \{6(n-2) + 3(n-2)(n-3) + 1\}$$
$$= 6(n-2) : \{3(n-1)(n-2) + 1\} \dots\dots\dots\dots(\beta).$$

Combining (α) and (β), we obtain
$$R_n : R_{n-1} = (n-1) : (n-2).$$
Thus

$R_2, R_3, R_4, \ldots R_n$ are in the ratio of the successive numbers 1, 2, 3 ... $(n-1)$.

Proposition 28.

If O be the origin and BC any arc measured in the 'forward' direction on any turn of the spiral, let two circles be drawn (1) with centre O, and radius OB, meeting OC in C', and (2) with centre O and radius OC, meeting OB produced in B'. Then, if E denote the area bounded by the larger circular arc $B'C$, the line $B'B$, and the spiral BC, while F denotes the area bounded by the smaller arc BC', the line CC' and the spiral BC,
$$E : F = \{OB + \tfrac{2}{3}(OC - OB)\} : \{OB + \tfrac{1}{3}(OC - OB)\}.$$

Let σ denote the area of the lesser sector OBC'; then the larger sector $OB'C$ is equal to $\sigma + F + E$.

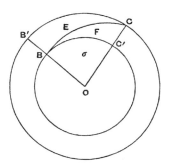

Thus [Prop. 26]
$(\sigma + F) : (\sigma + F + E) = \{OC \cdot OB + \tfrac{1}{3}(OC - OB)^2\} : OC^2 \ldots (1),$
whence
$$E : (\sigma + F) = \{OC(OC - OB) - \tfrac{1}{3}(OC - OB)^2\}$$
$$: \{OC \cdot OB + \tfrac{1}{3}(OC - OB)^2\}$$
$$= \{OB(OC - OB) + \tfrac{2}{3}(OC - OB)^2\}$$
$$: \{OC \cdot OB + \tfrac{1}{3}(OC - OB)^2\} \ldots \ldots \ldots (2).$$

Again
$$(\sigma + F + E) : \sigma = OC^2 : OB^2.$$
Therefore, by the first proportion above, *ex aequali*,
$$(\sigma + F) : \sigma = \{OC \cdot OB + \tfrac{1}{3}(OC - OB)^2\} : OB^2,$$
whence
$$(\sigma + F) : F = \{OC \cdot OB + \tfrac{1}{3}(OC - OB)^2\}$$
$$: \{OB(OC - OB) + \tfrac{1}{3}(OC - OB)^2\}.$$

Combining this with (2) above, we obtain
$$E : F = \{OB(OC - OB) + \tfrac{2}{3}(OC - OB)^2\}$$
$$: \{OB(OC - OB) + \tfrac{1}{3}(OC - OB)^2\}$$
$$= \{OB + \tfrac{2}{3}(OC - OB)\} : \{OB + \tfrac{1}{3}(OC - OB)\}.$$

ON THE EQUILIBRIUM OF PLANES

OR

THE CENTRES OF GRAVITY OF PLANES.

BOOK I.

"I POSTULATE the following:

1. Equal weights at equal distances are in equilibrium, and equal weights at unequal distances are not in equilibrium but incline towards the weight which is at the greater distance.

2. If, when weights at certain distances are in equilibrium, something be added to one of the weights, they are not in equilibrium but incline towards that weight to which the addition was made.

3. Similarly, if anything be taken away from one of the weights, they are not in equilibrium but incline towards the weight from which nothing was taken.

4. When equal and similar plane figures coincide if applied to one another, their centres of gravity similarly coincide.

5. In figures which are unequal but similar the centres of gravity will be similarly situated. By points similarly situated in relation to similar figures I mean points such that, if straight lines be drawn from them to the equal angles, they make equal angles with the corresponding sides.

6. If magnitudes at certain distances be in equilibrium, (other) magnitudes equal to them will also be in equilibrium at the same distances.

7. In any figure whose perimeter is concave in (one and) the same direction the centre of gravity must be within the figure."

Proposition 1.

Weights which balance at equal distances are equal.

For, if they are unequal, take away from the greater the difference between the two. The remainders will then not balance [*Post.* 3]; which is absurd.

Therefore the weights cannot be unequal.

Proposition 2.

Unequal weights at equal distances will not balance but will incline towards the greater weight.

For take away from the greater the difference between the two. The equal remainders will therefore balance [*Post.* 1]. Hence, if we add the difference again, the weights will not balance but incline towards the greater [*Post.* 2].

Proposition 3.

Unequal weights will balance at unequal distances, the greater weight being at the lesser distance.

Let A, B be two unequal weights (of which A is the greater) balancing about C at distances AC, BC respectively.

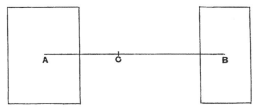

Then shall AC be less than BC. For, if not, take away from A the weight $(A-B.)$ The remainders will then incline towards B [*Post.* 3]. But this is impossible, for (1) if $AC = CB$, the equal remainders will balance, or (2) if $AC > CB$, they will incline towards A at the greater distance [*Post.* 1].

Hence $AC < CB$.

Conversely, if the weights balance, and $AC < CB$, then $A > B$.

Proposition 4.

If two equal weights have not the same centre of gravity, the centre of gravity of both taken together is at the middle point of the line joining their centres of gravity.

[Proved from Prop. 3 by *reductio ad absurdum*. Archimedes assumes that the centre of gravity of both together is on the straight line joining the centres of gravity of each, saying that this had been proved before ($προδέδεικται$). The allusion is no doubt to the lost treatise *On levers* ($περὶ ζυγῶν$).]

Proposition 5.

If three equal magnitudes have their centres of gravity on a straight line at equal distances, the centre of gravity of the system will coincide with that of the middle magnitude.

[This follows immediately from Prop. 4.]

COR 1. *The same is true of any odd number of magnitudes if those which are at equal distances from the middle one are equal, while the distances between their centres of gravity are equal.*

COR. 2. *If there be an even number of magnitudes with their centres of gravity situated at equal distances on one straight line, and if the two middle ones be equal, while those which are equidistant from them (on each side) are equal respectively, the centre of gravity of the system is the middle point of the line joining the centres of gravity of the two middle ones.*

Propositions 6, 7.

Two magnitudes, whether commensurable [Prop. 6] *or incommensurable* [Prop. 7], *balance at distances reciprocally proportional to the magnitudes.*

I. Suppose the magnitudes A, B to be commensurable, and the points A, B to be their centres of gravity. Let DE be a straight line so divided at C that

$$A : B = DC : CE.$$

We have then to prove that, if A be placed at E and B at D, C is the centre of gravity of the two taken together.

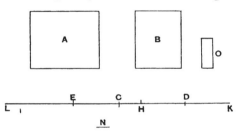

Since A, B are commensurable, so are DC, CE. Let N be a common measure of DC, CE. Make DH, DK each equal to CE, and EL (on CE produced) equal to CD. Then $EH = CD$, since $DH = CE$. Therefore LH is bisected at E, as HK is bisected at D.

Thus LH, HK must each contain N an even number of times.

Take a magnitude O such that O is contained as many times in A as N is contained in LH, whence

$$A : O = LH : N.$$

But $$B : A = CE : DC$$
$$= HK : LH.$$

Hence, *ex aequali*, $B : O = HK : N$, or O is contained in B as many times as N is contained in HK.

Thus O is a common measure of A, B.

Divide LH, HK into parts each equal to N, and A, B into parts each equal to O. The parts of A will therefore be equal in number to those of LH, and the parts of B equal in number to those of HK. Place one of the parts of A at the middle point of each of the parts N of LH, and one of the parts of B at the middle point of each of the parts N of HK.

Then the centre of gravity of the parts of A placed at equal distances on LH will be at E, the middle point of LH [Prop. 5, Cor. 2], and the centre of gravity of the parts of B placed at equal distances along HK will be at D, the middle point of HK.

Thus we may suppose A itself applied at E, and B itself applied at D.

But the system formed by the parts O of A and B together is a system of equal magnitudes even in number and placed at equal distances along LK. And, since $LE = CD$, and $EC = DK$, $LC = CK$, so that C is the middle point of LK. Therefore C is the centre of gravity of the system ranged along LK.

Therefore A acting at E and B acting at D balance about the point C.

II. Suppose the magnitudes to be incommensurable, and let them be $(A + a)$ and B respectively. Let DE be a line divided at C so that
$$(A + a) : B = DC : CE.$$

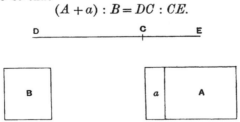

Then, if $(A + a)$ placed at E and B placed at D do not balance about C, $(A + a)$ is either too great to balance B, or not great enough.

Suppose, if possible, that $(A + a)$ is too great to balance B. Take from $(A + a)$ a magnitude a smaller than the deduction which would make the remainder balance B, but such that the remainder A and the magnitude B are commensurable.

Then, since A, B are commensurable, and
$$A : B < DC : CE,$$
A and B will not balance [Prop. 6], but D will be depressed.

But this is impossible, since the deduction a was an insufficient deduction from $(A + a)$ to produce equilibrium, so that E was still depressed.

Therefore $(A + a)$ is not too great to balance B; and similarly it may be proved that B is not too great to balance $(A + a)$.

Hence $(A + a)$, B taken together have their centre of gravity at C.

Proposition 8.

If AB be a magnitude whose centre of gravity is C, and AD a part of it whose centre of gravity is F, then the centre of gravity of the remaining part will be a point G on FC produced such that
$$GC : CF = (AD) : (DE).$$

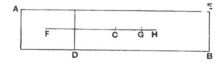

For, if the centre of gravity of the remainder (DE) be not G, let it be a point H. Then an absurdity follows at once from Props. 6, 7.

Proposition 9.

The centre of gravity of any parallelogram lies on the straight line joining the middle points of opposite sides.

Let $ABCD$ be a parallelogram, and let EF join the middle points of the opposite sides AD, BC.

If the centre of gravity does not lie on EF, suppose it to be H, and draw HK parallel to AD or BC meeting EF in K.

Then it is possible, by bisecting *ED*, then bisecting the halves, and so on continually, to arrive at a length *EL* less

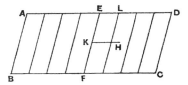

than *KH*. Divide both *AE* and *ED* into parts each equal to *EL*, and through the points of division draw parallels to *AB* or *CD*.

We have then a number of equal and similar parallelograms, and, if any one be applied to any other, their centres of gravity coincide [*Post.* 4]. Thus we have an even number of equal magnitudes whose centres of gravity lie at equal distances along a straight line. Hence the centre of gravity of the whole parallelogram will lie on the line joining the centres of gravity of the two middle parallelograms [Prop. 5, Cor. 2].

But this is impossible, for *H* is outside the middle parallelograms.

Therefore the centre of gravity cannot but lie on *EF*.

Proposition 10.

The centre of gravity of a parallelogram is the point of intersection of its diagonals.

For, by the last proposition, the centre of gravity lies on each of the lines which bisect opposite sides. Therefore it is at the point of their intersection; and this is also the point of intersection of the diagonals.

Alternative proof.

Let *ABCD* be the given parallelogram, and *BD* a diagonal. Then the triangles *ABD*, *CDB* are equal and similar, so that [*Post.* 4], if one be applied to the other, their centres of gravity will fall one upon the other.

Suppose F to be the centre of gravity of the triangle ABD.
Let G be the middle point of BD.
Join FG and produce it to H, so that $FG = GH$.

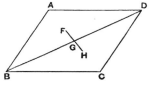

If we then apply the triangle ABD to the triangle CDB so that AD falls on CB and AB on CD, the point F will fall on H.

But [by *Post.* 4] F will fall on the centre of gravity of CDB. Therefore H is the centre of gravity of CDB.

Hence, since F, H are the centres of gravity of the two equal triangles, the centre of gravity of the whole parallelogram is at the middle point of FH, i.e. at the middle point of BD, which is the intersection of the two diagonals.

Proposition 11.

If abc, ABC be two similar triangles, and g, G two points in them similarly situated with respect to them respectively, then, if g be the centre of gravity of the triangle abc, G must be the centre of gravity of the triangle ABC.

Suppose $ab : bc : ca = AB : BC : CA$.

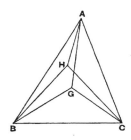

The proposition is proved by an obvious *reductio ad absurdum*. For, if G be not the centre of gravity of the triangle ABC, suppose H to be its centre of gravity.

Post. 5 requires that g, H shall be similarly situated with respect to the triangles respectively; and this leads at once to the absurdity that the angles HAB, GAB are equal.

Proposition 12.

Given two similar triangles abc, ABC, and d, D the middle points of bc, BC respectively, then, if the centre of gravity of abc lie on ad, that of ABC will lie on AD.

Let g be the point on ad which is the centre of gravity of abc.

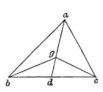 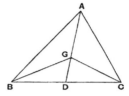

Take G on AD such that
$$ad : ag = AD : AG,$$
and join gb, gc, GB, GC.

Then, since the triangles are similar, and bd, BD are the halves of bc, BC respectively,
$$ab : bd = AB : BD,$$
and the angles abd, ABD are equal.

Therefore the triangles abd, ABD are similar, and
$$\angle bad = \angle BAD.$$
Also $\qquad ba : ad = BA : AD,$
while, from above, $\quad ad : ag = AD : AG.$

Therefore $ba : ag = BA : AG$, while the angles bag, BAG are equal.

Hence the triangles bag, BAG are similar, and
$$\angle abg = \angle ABG.$$

And, since the angles abd, ABD are equal, it follows that
$$\angle gbd = \angle GBD.$$
In exactly the same manner we prove that
$$\angle gac = \angle GAC,$$
$$\angle acg = \angle ACG,$$
$$\angle gcd = \angle GCD.$$

Therefore g, G are similarly situated with respect to the triangles respectively; whence [Prop. 11] G is the centre of gravity of ABC.

Proposition 13.

In any triangle the centre of gravity lies on the straight line joining any angle to the middle point of the opposite side.

Let ABC be a triangle and D the middle point of BC. Join AD. Then shall the centre of gravity lie on AD.

For, if possible, let this not be the case, and let H be the centre of gravity. Draw HI parallel to CB meeting AD in I.

Then, if we bisect DC, then bisect the halves, and so on, we shall at length arrive at a length, as DE, less than HI.

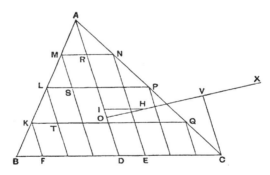

Divide both BD and DC into lengths each equal to DE, and through the points of division draw lines each parallel to DA meeting BA and AC in points as K, L, M and N, P, Q respectively.

Join MN, LP, KQ, which lines will then be each parallel to BC.

We have now a series of parallelograms as FQ, TP, SN, and AD bisects opposite sides in each. Thus the centre of gravity of each parallelogram lies on AD [Prop. 9], and therefore the centre of gravity of the figure made up of them all lies on AD.

Let the centre of gravity of all the parallelograms taken together be O. Join OH and produce it; also draw CV parallel to DA meeting OH produced in V.

Now, if n be the number of parts into which AC is divided,

$\triangle ADC$: (sum of triangles on $AN, NP, \ldots$)
$$= AC^2 : (AN^2 + NP^2 + \ldots)$$
$$= n^2 : n$$
$$= n : 1$$
$$= AC : AN.$$

Similarly

$\triangle ABD$: (sum of triangles on $AM, ML, \ldots$) $= AB : AM$.

And $\qquad AC : AN = AB : AM$.

It follows that

$\triangle ABC$: (sum of all the small $\triangle$s) $= CA : AN$
$$> VO : OH, \text{ by parallels.}$$

Suppose OV produced to X so that

$\triangle ABC$: (sum of small $\triangle$s) $= XO : OH$,

whence, *dividendo*,

(sum of parallelograms) : (sum of small $\triangle$s) $= XH : HO$.

Since then the centre of gravity of the triangle ABC is at H, and the centre of gravity of the part of it made up of the parallelograms is at O, it follows from Prop. 8 that the centre of gravity of the remaining portion consisting of all the small triangles taken together is at X.

But this is impossible, since all the triangles are on one side of the line through X parallel to AD.

Therefore the centre of gravity of the triangle cannot but lie on AD.

Alternative proof.

Suppose, if possible, that H, not lying on AD, is the centre of gravity of the triangle ABC. Join AH, BH, CH. Let E, F be the middle points of CA, AB respectively, and join DE, EF, FD. Let EF meet AD in M.

Draw FK, EL parallel to AH meeting BH, CH in K, L respectively. Join KD, HD, LD, KL. Let KL meet DH in N, and join MN.

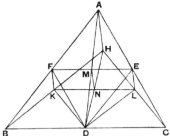

Since DE is parallel to AB, the triangles ABC, EDC are similar.

And, since $CE = EA$, and EL is parallel to AH, it follows that $CL = LH$. And $CD = DB$. Therefore BH is parallel to DL.

Thus in the similar and similarly situated triangles ABC, EDC the straight lines AH, BH are respectively parallel to EL, DL; and it follows that H, L are similarly situated with respect to the triangles respectively.

But H is, by hypothesis, the centre of gravity of ABC. Therefore L is the centre of gravity of EDC. [Prop. 11]

Similarly the point K is the centre of gravity of the triangle FBD.

And the triangles FBD, EDC are equal, so that the centre of gravity of both together is at the middle point of KL, i.e. at the point N.

The remainder of the triangle ABC, after the triangles FBD, EDC are deducted, is the parallelogram $AFDE$, and the centre of gravity of this parallelogram is at M, the intersection of its diagonals.

It follows that the centre of gravity of the whole triangle ABC must lie on MN; that is, MN must pass through H, which is impossible (since MN is parallel to AH).

Therefore the centre of gravity of the triangle ABC cannot but lie on AD.

Proposition 14.

It follows at once from the last proposition that *the centre of gravity of any triangle is at the intersection of the lines drawn from any two angles to the middle points of the opposite sides respectively.*

Proposition 15.

If AD, BC be the two parallel sides of a trapezium $ABCD$, AD being the smaller, and if AD, BC be bisected at E, F respectively, then the centre of gravity of the trapezium is at a point G on EF such that

$$GE : GF = (2BC + AD) : (2AD + BC).$$

Produce BA, CD to meet at O. Then FE produced will also pass through O, since $AE = ED$, and $BF = FC$.

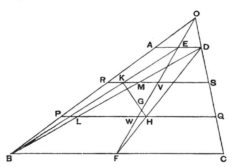

Now the centre of gravity of the triangle OAD will lie on OE, and that of the triangle OBC will lie on OF. [Prop. 13]

It follows that the centre of gravity of the remainder, the trapezium $ABCD$, will also lie on OF. [Prop. 8]

Join BD, and divide it at L, M into three equal parts. Through L, M draw PQ, RS parallel to BC meeting BA in P, R, FE in W, V, and CD in Q, S respectively.

Join DF, BE meeting PQ in H and RS in K respectively.

Now, since $\qquad BL = \tfrac{1}{3} BD,$

$\qquad\qquad\qquad FH = \tfrac{1}{3} FD.$

Therefore H is the centre of gravity of the triangle DBC*.

Similarly, since $EK = \frac{1}{3} BE$, it follows that K is the centre of gravity of the triangle ADB.

Therefore the centre of gravity of the triangles DBC, ADB together, i.e. of the trapezium, lies on the line HK.

But it also lies on OF.

Therefore, if OF, HK meet in G, G is the centre of gravity of the trapezium.

Hence [Props. 6, 7]
$$\triangle DBC : \triangle ABD = KG : GH$$
$$= VG : GW.$$
But $\triangle DBC : \triangle ABD = BC : AD.$

Therefore $BC : AD = VG : GW.$

It follows that
$$(2BC + AD) : (2AD + BC) = (2VG + GW) : (2GW + VG)$$
$$= EG : GF.$$
Q. E. D.

---

* This easy deduction from Prop. 14 is assumed by Archimedes without proof.

ON THE EQUILIBRIUM OF PLANES.

BOOK II.

Proposition 1.

If P, P' be two parabolic segments and D, E their centres of gravity respectively, the centre of gravity of the two segments taken together will be at a point C on DE determined by the relation

$$P : P' = CE : CD*.$$

In the same straight line with DE measure EH, EL each equal to DC, and DK equal to DH; whence it follows at once that $DK = CE$, and also that $KC = CL$.

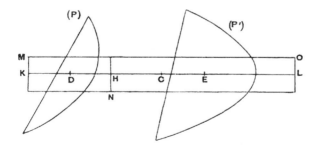

* This proposition is really a particular case of Props. 6, 7 of Book I. and is therefore hardly necessary. As, however, Book II. relates exclusively to parabolic segments, Archimedes' object was perhaps to emphasize the fact that the magnitudes in I. 6, 7 might be parabolic segments as well as rectilinear figures. His procedure is to substitute for the segments rectangles of equal area, a substitution which is rendered possible by the results obtained in his separate treatise on the *Quadrature of the Parabola*.

Apply a rectangle MN equal in area to the parabolic segment P to a base equal to KH, and place the rectangle so that KH bisects it, and is parallel to its base.

Then D is the centre of gravity of MN, since $KD = DH$.

Produce the sides of the rectangle which are parallel to KH, and complete the rectangle NO whose base is equal to HL. Then E is the centre of gravity of the rectangle NO.

Now
$$(MN):(NO) = KH : HL$$
$$= DH : EH$$
$$= CE : CD$$
$$= P : P'.$$
But $(MN) = P.$

Therefore $(NO) = P'.$

Also, since C is the middle point of KL, C is the centre of gravity of the whole parallelogram made up of the two parallelograms (MN), (NO), which are equal to, and have the same centres of gravity as, P, P' respectively.

Hence C is the centre of gravity of P, P' taken together.

Definition and lemmas preliminary to Proposition 2.

"If in a segment bounded by a straight line and a section of a right-angled cone [a parabola] a triangle be inscribed having the same base as the segment and equal height, if again triangles be inscribed in the remaining segments having the same bases as the segments and equal height, and if in the remaining segments triangles be inscribed in the same manner, let the resulting figure be said to be **inscribed in the recognised manner** (γνωρίμως ἐγγράφεσθαι) in the segment.

And it is plain

(1) that *the lines joining the two angles of the figure so inscribed which are nearest to the vertex of the segment, and the next*

ON THE EQUILIBRIUM OF PLANES II. 205

pairs of angles in order, will be parallel to the base of the segment,

(2) that *the said lines will be bisected by the diameter of the segment,* and

(3) that *they will cut the diameter in the proportions of the successive odd numbers, the number one having reference to* [*the length adjacent to*] *the vertex of the segment.*

And these properties will have to be proved in their proper places (ἐν ταῖς τάξεσιν)."

[The last words indicate an intention to give these propositions in their proper connexion with systematic proofs; but the intention does not appear to have been carried out, or at least we know of no lost work of Archimedes in which they could have appeared. The results can however be easily derived from propositions given in the *Quadrature of the Parabola* as follows.

(1) Let $BRQPApqrb$ be a figure inscribed 'in the recognised manner' in the parabolic segment BAb of which Bb is the base, A the vertex and AO the diameter.

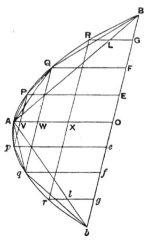

Bisect each of the lines BQ, BA, QA, Aq, Ab, qb, and through the middle points draw lines parallel to AO meeting Bb in G, F, E, e, f, g respectively.

These lines will then pass through the vertices R, Q, P, p, q, r of the respective parabolic segments [*Quadrature of the Parabola*, Prop. 18], i.e. through the angular points of the inscribed figure (since the triangles and segments are of equal height).

Also $BG = GF = FE = EO$, and $Oe = ef = fg = gb$. But $BO = Ob$, and therefore all the parts into which Bb is divided are equal.

If now AB, RG meet in L, and Ab, rg in l, we have

$$BG : GL = BO : OA, \text{ by parallels,}$$
$$= bO : OA$$
$$= bg : gl,$$

whence $GL = gl$.

Again [*ibid.*, Prop. 4]

$$GL : LR = BO : OG$$
$$= bO : Og$$
$$= gl : lr;$$

and, since $GL = gl$, $LR = lr$.

Therefore GR, gr are equal as well as parallel.

Hence $GRrg$ is a parallelogram, and Rr is parallel to Bb.

Similarly it may be shown that Pp, Qq are each parallel to Bb.

(2) Since $RGgr$ is a parallelogram, and RG, rg are parallel to AO, while $GO = Og$, it follows that Rr is bisected by AO.

And similarly for Pp, Qq.

(3) Lastly, if V, W, X be the points of bisection of Pp, Qq, Rr,

$$AV : AW : AX : AO = PV^2 : QW^2 : RX^2 : BO^2$$
$$= 1 : 4 : 9 : 16,$$

whence $AV : VW : WX : XO = 1 : 3 : 5 : 7.$]

Proposition 2.

If a figure be 'inscribed in the recognised manner' in a parabolic segment, the centre of gravity of the figure so inscribed will lie on the diameter of the segment.

For, in the figure of the foregoing lemmas, the centre of gravity of the trapezium $BRrb$ must lie on XO, that of the trapezium $RQqr$ on WX, and so on, while the centre of gravity of the triangle PAp lies on AV.

Hence the centre of gravity of the whole figure lies on AO.

Proposition 3.

If BAB', bab' be two similar parabolic segments whose diameters are AO, ao respectively, and if a figure be inscribed in each segment 'in the recognised manner,' the number of sides in each figure being equal, the centres of gravity of the inscribed figures will divide AO, ao in the same ratio.

[Archimedes enunciates this proposition as true of *similar* segments, but it is equally true of segments which are not similar, as the course of the proof will show.]

Suppose $BRQPAP'Q'R'B'$, $brqpap'q'r'b'$ to be the two figures inscribed 'in the recognised manner.' Join PP', QQ', RR' meeting AO in L, M, N, and pp', qq', rr' meeting ao in l, m, n.

Then [Lemma (3)]
$$AL : LM : MN : NO$$
$$= 1 : 3 : 5 : 7$$
$$= al : lm : mn : no,$$

so that AO, ao are divided in the same proportion.

Also, by reversing the proof of Lemma (3), we see that
$$PP' : pp' = QQ' : qq' = RR' : rr' = BB' : bb'.$$

Since then $RR' : BB' = rr' : bb'$, and these ratios respectively determine the proportion in which NO, no are divided by the centres of gravity of the trapezia $BRR'B'$, $brr'b'$ [I. 15], it follows that the centres of gravity of the trapezia divide NO, no in the same ratio.

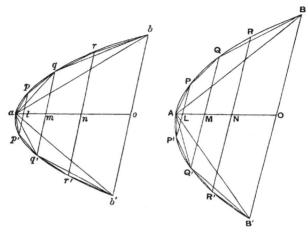

Similarly the centres of gravity of the trapezia $RQQ'R'$, $rqq'r'$ divide MN, mn in the same ratio respectively, and so on.

Lastly, the centres of gravity of the triangles PAP', pap' divide AL, al respectively in the same ratio.

Moreover the corresponding trapezia and triangles are, each to each, in the same proportion (since their sides and heights are respectively proportional), while AO, ao are divided in the same proportion.

Therefore the centres of gravity of the complete inscribed figures divide AO, ao in the same proportion.

Proposition 4.

The centre of gravity of any parabolic segment cut off by a straight line lies on the diameter of the segment.

Let BAB' be a parabolic segment, A its vertex and AO its diameter.

Then, if the centre of gravity of the segment does not lie on AO, suppose it to be, if possible, the point F. Draw FE parallel to AO meeting BB' in E.

Inscribe in the segment the triangle ABB' having the same vertex and height as the segment, and take an area S such that
$$\triangle ABB' : S = BE : EO.$$

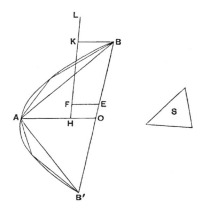

We can then inscribe in the segment 'in the recognised manner' a figure such that the segments of the parabola left over are together less than S. [For Prop. 20 of the *Quadrature of the Parabola* proves that, if in any segment the triangle with the same base and height be inscribed, the triangle is greater than half the segment; whence it appears that, each time that we increase the number of the sides of the figure inscribed 'in the recognised manner,' we take away more than half of the remaining segments.]

Let the inscribed figure be drawn accordingly; its centre of gravity then lies on AO [Prop. 2]. Let it be the point H.

Join HF and produce it to meet in K the line through B parallel to AO.

Then we have

(inscribed figure) : (remainder of segmt.) $> \triangle ABB' : S$
$> BE : EO$
$> KF : FH.$

Suppose L taken on HK produced so that the former ratio is equal to the ratio $LF : FH$.

Then, since *H* is the centre of gravity of the inscribed figure, and *F* that of the segment, *L* must be the centre of gravity of all the segments taken together which form the remainder of the original segment. [I. 8]

But this is impossible, since all these segments lie on one side of the line drawn through *L* parallel to *AO* [Cf. *Post.* 7].

Hence the centre of gravity of the segment cannot but lie on *AO*.

Proposition 5.

If in a parabolic segment a figure be inscribed 'in the recognised manner,' the centre of gravity of the segment is nearer to the vertex of the segment than the centre of gravity of the inscribed figure is.

Let *BAB'* be the given segment, and *AO* its diameter. *First*, let *ABB'* be the *triangle* inscribed 'in the recognised manner.'

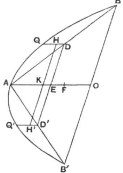

Divide *AO* in *F* so that $AF = 2FO$; *F* is then the centre of gravity of the triangle *ABB'*.

Bisect *AB*, *AB'* in *D*, *D'* respectively, and join *DD'* meeting *AO* in *E*. Draw *DQ*, *D'Q'* parallel to *OA* to meet the curve. *QD*, *Q'D'* will then be the diameters of the segments whose bases are *AB*, *AB'*, and the centres of gravity of those segments will lie respectively on *QD*, *Q'D'* [Prop. 4]. Let them be *H*, *H'*, and join *HH'* meeting *AO* in *K*.

Now *QD*, *Q'D'* are equal*, and therefore the segments of which they are the diameters are equal [*On Conoids and Spheroids*, Prop. 3].

---

* This may either be inferred from Lemma (1) above (since *QQ'*, *DD'* are both parallel to *BB'*), or from Prop. 19 of the *Quadrature of the Parabola*, which applies equally to *Q* or *Q'*.

ON THE EQUILIBRIUM OF PLANES II.

Also, since $QD, Q'D'$ are parallel*, and $DE = ED'$, K is the middle point of HH'.

Hence the centre of gravity of the equal segments AQB, $AQ'B'$ taken together is K, where K lies between E and A. And the centre of gravity of the triangle ABB' is F.

It follows that the centre of gravity of the whole segment BAB' lies between K and F, and is therefore nearer to the vertex A than F is.

Secondly, take the *five-sided* figure $BQAQ'B'$ inscribed 'in the recognised manner,' $QD, Q'D'$ being, as before, the diameters of the segments $AQB, AQ'B'$.

Then, by the first part of this proposition, the centre of gravity of the segment AQB (lying of course on QD) is nearer to Q than the centre of gravity of the triangle AQB is. Let the centre of gravity of the segment be H, and that of the triangle I.

Similarly let H' be the centre of gravity of the segment $AQ'B'$, and I' that of the triangle $AQ'B'$.

It follows that the centre of gravity of the two segments $AQB, AQ'B'$ taken together is K, the middle point of HH', and that of the two triangles $AQB, AQ'B'$ is L, the middle point of II'.

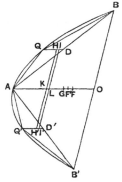

If now the centre of gravity of the triangle ABB' be F, the centre of gravity of the whole segment BAB' (i.e. that of the triangle ABB' and the two segments $AQB, AQ'B'$ taken together) is a point G on KF determined by the proportion

(sum of segments $AQB, AQ'B'$) : $\triangle ABB' = FG : GK$. [I. 6, 7]

* There is clearly some interpolation in the text here, which has the words καὶ ἐπεὶ παραλληλόγραμμόν ἐστι τὸ ΘΖΗΙ. It is not yet proved that $H'D'DH$ is a *parallelogram*; this can only be inferred from the fact that H, H' divide QD, $Q'D'$ respectively in the same ratio. But this latter property does not appear till Prop. 7, and is then only enunciated of *similar* segments. The interpolation must have been made before Eutocius' time, because he has a note on the phrase, and explains it by gravely assuming that H, H' divide $QD, Q'D'$ respectively in the same ratio.

And the centre of gravity of the inscribed figure $BQAQ'B'$ is a point F' on LF determined by the proportion

$$(\triangle AQB + \triangle AQ'B') : \triangle ABB' = FF' : F'L. \quad \text{[I. 6, 7]}$$

[Hence $\qquad FG : GK > FF' : F'L$,

or $\qquad GK : FG < F'L : FF'$,

and, *componendo*, $\quad FK : FG < FL : FF'$, while $FK > FL$.]

Therefore $FG > FF'$, or G lies nearer than F' to the vertex A.

Using this last result, and proceeding in the same way, we can prove the proposition for *any* figure inscribed 'in the recognised manner.'

Proposition 6.

Given a segment of a parabola cut off by a straight line, it is possible to inscribe in it 'in the recognised manner' a figure such that the distance between the centres of gravity of the segment and of the inscribed figure is less than any assigned length.

Let BAB' be the segment, AO its diameter, G its centre of gravity, and ABB' the triangle inscribed 'in the recognised manner.'

Let D be the assigned length and S an area such that

$$AG : D = \triangle ABB' : S.$$

In the segment inscribe 'in the recognised manner' a figure such that the sum of the segments left over is less than S. Let F be the centre of gravity of the inscribed figure.

We shall prove that $FG < D$.

For, if not, FG must be either equal to, or greater than, D.

And clearly

$\qquad$(inscribed fig.) : (sum of remaining segmts.)

$\qquad\qquad > \triangle ABB' : S$

$\qquad\qquad > AG : D$

$\qquad\qquad > AG : FG$, by hypothesis (since $FG \not< D$).

Let the first ratio be equal to the ratio $KG : FG$ (where K lies on GA produced); and it follows that K is the centre of gravity of the small segments taken together. [I. 8]

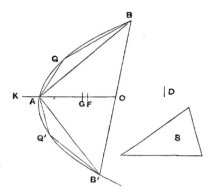

But this is impossible, since the segments are all on the same side of a line drawn through K parallel to BB'.

Hence FG cannot but be less than D.

Proposition 7.

If there be two similar parabolic segments, their centres of gravity divide their diameters in the same ratio.

[This proposition, though enunciated of *similar* segments only, like Prop. 3 on which it depends, is equally true of *any* segments. This fact did not escape Archimedes, who uses the proposition in its more general form for the proof of Prop. 8 immediately following.]

Let BAB', bab' be the two similar segments, AO, ao their diameters, and G, g their centres of gravity respectively.

Then, if G, g do not divide AO, ao respectively in the same ratio, suppose H to be such a point on AO that

$$AH : HO = ag : go\,;$$

and inscribe in the segment BAB' 'in the recognised manner' a figure such that, if F be its centre of gravity,

$$GF < GH. \qquad \text{[Prop. 6]}$$

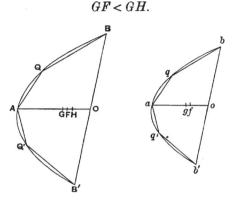

Inscribe in the segment bab' 'in the recognised manner' a similar figure; then, if f be the centre of gravity of this figure,

$$ag < af. \qquad \text{[Prop. 5]}$$

And, by Prop. 3, $\quad af : fo = AF : FO$.

But $\qquad AF : FO < AH : HO$

$\qquad\qquad\qquad < ag : go$, by hypothesis.

Therefore $\qquad af : fo < ag : go$; which is impossible.

It follows that G, g cannot but divide AO, ao in the same ratio.

Proposition 8.

If AO be the diameter of a parabolic segment, and G its centre of gravity, then

$$AG = \tfrac{3}{2} GO.$$

Let the segment be BAB'. Inscribe the triangle ABB' 'in the recognised manner,' and let F be its centre of gravity.

Bisect AB, AB' in D, D', and draw $DQ, D'Q'$ parallel to OA to meet the curve, so that $QD, Q'D'$ are the diameters of the segments $AQB, AQ'B'$ respectively.

Let H, H' be the centres of gravity of the segments $AQB, AQ'B'$ respectively. Join QQ', HH' meeting AO in V, K respectively.

K is then the centre of gravity of the two segments AQB, $AQ'B'$ taken together.

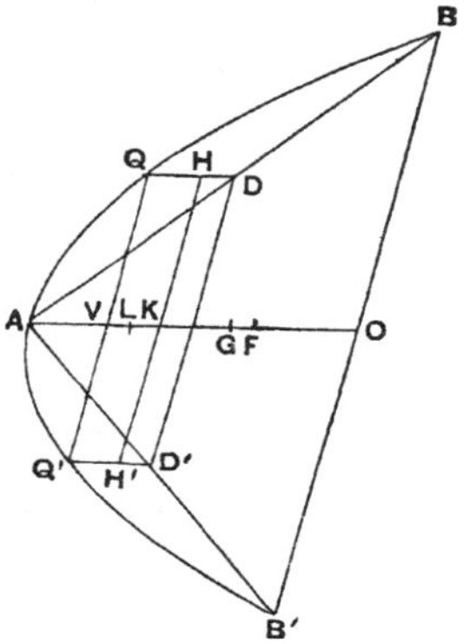

Now $\quad AG : GO = QH : HD,$

[Prop. 7]

whence $\quad AO : OG = QD : HD.$

But $AO = 4QD$ [as is easily proved by means of Lemma (3), p. 206].

Therefore $\quad OG = 4HD$;

and, by subtraction, $\quad AG = 4QH.$

Also, by Lemma (2), QQ' is parallel to BB' and therefore to DD'. It follows from Prop. 7 that HH' is also parallel to QQ' or DD',

and hence $\quad QH = VK.$

Therefore $\quad AG = 4VK,$

and $\quad AV + KG = 3VK.$

Measuring VL along VK so that $VL = \tfrac{1}{3} AV$, we have

$$KG = 3LK \dots\dots\dots\dots\dots\dots\dots(1).$$

Again $\quad AO = 4AV \quad\quad$ [Lemma (3)]

$\quad\quad\quad\quad = 3AL,$ since $AV = 3VL,$

whence $\quad AL = \tfrac{1}{3} AO = OF \dots\dots\dots\dots (2).$

Now, by I. 6, 7,

$\triangle ABB'$: (sum of segmts. $AQB, AQ'B') = KG : GF,$

and $\quad \triangle ABB' = 3$ (sum of segments $AQB, AQ'B'$)

[since the segment ABB' is equal to $\tfrac{4}{3} \triangle ABB'$ (*Quadrature of the Parabola*, Props. 17, 24)].

Hence $\quad KG = 3GF.$

But $\quad KG = 3LK,$ from (1) above.

Therefore $\quad LF = LK + KG + GF$

$\quad\quad\quad\quad = 5GF.$

And, from (2),
$$LF = (AO - AL - OF) = \tfrac{1}{3} AO = OF.$$
Therefore $OF = 5GF,$
and $OG = 6GF.$
But $AO = 3OF = 15GF.$
Therefore, by subtraction,
$$AG = 9GF$$
$$= \tfrac{3}{2} GO.$$

Proposition 9 (Lemma).

If a, b, c, d be four lines in continued proportion and in descending order of magnitude, and if
$$d : (a-d) = x : \tfrac{3}{5}(a-c),$$
and $(2a + 4b + 6c + 3d) : (5a + 10b + 10c + 5d) = y : (a-c),$
it is required to prove that
$$x + y = \tfrac{2}{3} a.$$

[The following is the proof given by Archimedes, with the only difference that it is set out in algebraical instead of geometrical notation. This is done in the particular case simply in order to make the proof easier to follow. Archimedes exhibits his lines in the figure reproduced in the margin, but, now that it is possible to use algebraical notation, there is no advantage in using the figure and the more cumbrous notation which only obscures the course of the proof. The relation between Archimedes' figure and the letters used below is as follows;

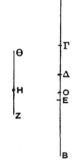

$AB = a$, $\Gamma B = b$, $\Delta B = c$, $EB = d$, $ZH = x$, $H\Theta = y$, $\Delta O = z$.]

We have
$$\frac{a}{b} = \frac{b}{c} = \frac{c}{d} \dots\dots\dots\dots\dots\dots\dots\dots (1),$$

whence
$$\frac{a-b}{b} = \frac{b-c}{c} = \frac{c-d}{d},$$

and therefore
$$\frac{a-b}{b-c} = \frac{b-c}{c-d} = \frac{a}{b} = \frac{b}{c} = \frac{c}{d} \dots\dots\dots (2).$$

Now $\dfrac{2(a+b)}{2c} = \dfrac{a+b}{c} = \dfrac{a+b}{b} \cdot \dfrac{b}{c} = \dfrac{a-c}{b-c} \cdot \dfrac{b-c}{c-d} = \dfrac{a-c}{c-d}.$

And, in like manner,
$$\frac{b+c}{d} = \frac{b+c}{c} \cdot \frac{c}{d} = \frac{a-c}{c-d}.$$

It follows from the last two relations that
$$\frac{a-c}{c-d} = \frac{2a+3b+c}{2c+d} \quad\ldots\ldots\ldots\ldots (3).$$

Suppose z to be so taken that
$$\frac{2a+4b+4c+2d}{2c+d} = \frac{a-c}{z} \quad\ldots\ldots\ldots (4),$$
so that $z < (c-d)$.

Therefore $\quad \dfrac{a-c+z}{a-c} = \dfrac{2a+4b+6c+3d}{2(a+d)+4(b+c)}.$

And, by hypothesis,
$$\frac{a-c}{y} = \frac{5(a+d)+10(b+c)}{2a+4b+6c+3d},$$
so that $\quad \dfrac{a-c+z}{y} = \dfrac{5(a+d)+10(b+c)}{2(a+d)+4(b+c)} = \dfrac{5}{2} \,\ldots\ldots\, (5).$

Again, dividing (3) by (4) crosswise, we obtain
$$\frac{z}{c-d} = \frac{2a+3b+c}{2(a+d)+4(b+c)},$$
whence $\quad \dfrac{c-d-z}{c-d} = \dfrac{b+3c+2d}{2(a+d)+4(b+c)} \quad\ldots\ldots\ldots (6).$

But, by (2),
$$\frac{c-d}{d} = \frac{a-b}{b} = \frac{3(b-c)}{3c} = \frac{2(c-d)}{2d},$$
so that $\quad \dfrac{c-d}{d} = \dfrac{(a-b)+3(b-c)+2(c-d)}{b+3c+2d} \quad\ldots\ldots (7).$

Combining (6) and (7), we have
$$\frac{c-d-z}{d} = \frac{(a-b)+3(b-c)+2(c-d)}{2(a+d)+4(b+c)},$$
whence $\quad \dfrac{c-z}{d} = \dfrac{3a+6b+3c}{2(a+d)+4(b+c)} \quad\ldots\ldots\ldots (8).$

And, since [by (1)]
$$\frac{c-d}{c+d} = \frac{b-c}{b+c} = \frac{a-b}{a+b},$$

we have
$$\frac{c-d}{a-c} = \frac{c+d}{b+c+a+b},$$
whence
$$\frac{a-d}{a-c} = \frac{a+2b+2c+d}{a+2b+c} = \frac{2(a+d)+4(b+c)}{2(a+c)+4b} \quad\ldots\ldots(9).$$
Thus
$$\frac{a-d}{\tfrac{3}{5}(a-c)} = \frac{2(a+d)+4(b+c)}{\tfrac{3}{5}\{2(a+c)+4b\}},$$
and therefore, by hypothesis,
$$\frac{d}{x} = \frac{2(a+d)+4(b+c)}{\tfrac{3}{5}\{2(a+c)+4b\}}.$$
But, by (8),
$$\frac{c-z}{d} = \frac{3a+6b+3c}{2(a+d)+4(b+c)};$$
and it follows, *ex aequali*, that
$$\frac{c-z}{x} = \frac{3(a+c)+6b}{\tfrac{3}{5}\{2(a+c)+4b\}} = \frac{5}{3}\cdot\frac{3}{2} = \frac{5}{2}.$$
And, by (5),
$$\frac{a-c+z}{y} = \frac{5}{2}.$$
Therefore
$$\frac{5}{2} = \frac{a}{x+y},$$
or
$$x+y = \tfrac{2}{5}a.$$

Proposition 10.

If $PP'B'B$ be the portion of a parabola intercepted between two parallel chords PP', BB' bisected respectively in N, O by the diameter ANO (N being nearer than O to A, the vertex of the segments), and if NO be divided into five equal parts of which LM is the middle one (L being nearer than M to N), then, if G be a point on LM such that

$$LG : GM = BO^2.(2PN+BO) : PN^2.(2BO+PN),$$

G will be the centre of gravity of the area $PP'B'B$.

Take a line ao equal to AO, and an on it equal to AN. Let p, q be points on the line ao such that

$$ao : aq = aq : an \quad\ldots\ldots\ldots\ldots\ldots(1),$$
$$ao : an = aq : ap \quad\ldots\ldots\ldots\ldots\ldots(2),$$

[whence $ao : aq = aq : an = an : ap$, or ao, aq, an, ap are lines in continued proportion and in descending order of magnitude].

Measure along GA a length GF such that

$$op : ap = OL : GF \ldots\ldots\ldots\ldots\ldots(3).$$

ON THE EQUILIBRIUM OF PLANES II.

Then, since PN, BO are ordinates to ANO,
$$BO^2 : PN^2 = AO : AN$$
$$= ao : an$$
$$= ao^2 : aq^2, \text{ by (1)},$$
so that $\quad BO : PN = ao : aq \quad \ldots\ldots\ldots\ldots\ldots\ldots\ldots (4),$
and $\quad BO^3 : PN^3 = ao^3 : aq^3$
$$= (ao : aq).(aq : an).(an : ap)$$
$$= ao : ap \quad \ldots\ldots\ldots\ldots\ldots\ldots\ldots (5).$$

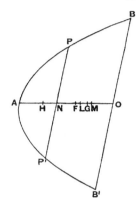

Thus (segment BAB') : (segment PAP')
$$= \triangle BAB' : \triangle PAP'$$
$$= BO^3 : PN^3$$
$$= ao : ap,$$
whence

(area $PP'B'B$) : (segment PAP') $= op : ap$
$$= OL : GF, \text{ by (3)},$$
$$= \tfrac{3}{5} ON : GF \ldots\ldots\ldots (6).$$
Now $\quad BO^2.(2PN + BO) : BO^3 = (2PN + BO) : BO$
$$= (2aq + ao) : ao, \text{ by (4)},$$
$$BO^3 : PN^3 = ao : ap, \text{ by (5)},$$
and $\quad PN^3 : PN^2.(2BO + PN) = PN : (2BO + PN)$
$$= aq : (2ao + aq), \text{ by (4)},$$
$$= ap : (2an + ap), \text{ by (2)}.$$

Hence, *ex aequali*,

$BO^2 \cdot (2PN + BO) : PN^2 \cdot (2BO + PN) = (2aq + ao) : (2an + ap)$,

so that, by hypothesis,

$$LG : GM = (2aq + ao) : (2an + ap).$$

Componendo, and multiplying the antecedents by 5,

$$ON : GM = \{5(ao + ap) + 10(aq + an)\} : (2an + ap).$$

But $ON : OM = 5 : 2$

$$= \{5(ao + ap) + 10(aq + an)\} : \{2(ao + ap) + 4(aq + an)\}.$$

It follows that

$ON : OG = \{5(ao + ap) + 10(aq + an)\} : (2ao + 4aq + 6an + 3ap)$.

Therefore

$(2ao + 4aq + 6an + 3ap) : \{5(ao + ap) + 10(aq + an)\} = OG : ON$
$$= OG : on.$$

And $\qquad ap : (ao - ap) = ap : op$
$$= GF : OL, \text{ by hypothesis,}$$
$$= GF : \tfrac{3}{5} on,$$

while ao, aq, an, ap are in continued proportion.

Therefore, by Prop. 9,

$$GF + OG = OF = \tfrac{2}{5} ao = \tfrac{2}{5} OA.$$

Thus F is the centre of gravity of the segment BAB'. [Prop. 8]

Let H be the centre of gravity of the segment PAP', so that $AH = \tfrac{3}{5} AN$.

And, since $\qquad AF = \tfrac{3}{5} AO$,

we have, by subtraction, $\quad HF = \tfrac{3}{5} ON$.

But, by (6) above,

$$(\text{area } PP'B'B) : (\text{segment } PAP') = \tfrac{3}{5} ON : GF$$
$$= HF : FG.$$

Thus, since F, H are the centres of gravity of the segments BAB', PAP' respectively, it follows [by I. 6, 7] that G is the centre of gravity of the area $PP'B'B$.

THE SAND-RECKONER.

"THERE are some, king Gelon, who think that the number of the sand is infinite in multitude; and I mean by the sand not only that which exists about Syracuse and the rest of Sicily but also that which is found in every region whether inhabited or uninhabited. Again there are some who, without regarding it as infinite, yet think that no number has been named which is great enough to exceed its multitude. And it is clear that they who hold this view, if they imagined a mass made up of sand in other respects as large as the mass of the earth, including in it all the seas and the hollows of the earth filled up to a height equal to that of the highest of the mountains, would be many times further still from recognising that any number could be expressed which exceeded the multitude of the sand so taken. But I will try to show you by means of geometrical proofs, which you will be able to follow, that, of the numbers named by me and given in the work which I sent to Zeuxippus, some exceed not only the number of the mass of sand equal in magnitude to the earth filled up in the way described, but also that of a mass equal in magnitude to the universe. Now you are aware that 'universe' is the name given by most astronomers to the sphere whose centre is the centre of the earth and whose radius is equal to the straight line between the centre of the sun and the centre of the earth. This is the common account (τὰ γραφόμενα), as you have heard from astronomers. But Aristarchus of Samos brought out a book consisting of some hypotheses, in which the premisses lead to the result that the universe is many times greater than that now so called. His hypotheses are that the fixed stars and the sun remain unmoved, that the earth revolves about the sun in the circumference of a circle, the sun lying in the middle of the orbit, and that the sphere of the fixed stars, situated about the same centre as the sun, is so great that the circle in which he supposes the earth to revolve bears such a proportion to the distance of the fixed stars as the centre of the sphere bears to its surface. Now it is easy to see that this is impossible; for, since the centre of the sphere has no magnitude, we cannot conceive it to bear any ratio whatever to the surface of the sphere. We must however take Aristarchus to mean this: since we conceive the earth to be, as it were, the centre of the universe, the ratio which the earth bears to what we describe as the 'universe' is the same as the ratio which the sphere containing the circle in which he supposes the earth to revolve bears to the sphere of the fixed stars. For he adapts the proofs of his results to a hypothesis of this kind, and in particular he appears to suppose the magnitude of the sphere in which he represents the earth as moving to be equal to what we call the 'universe.'

I say then that, even if a sphere were made up of the sand, as great as Aristarchus supposes the sphere of the fixed stars to be, I shall still prove that, of the numbers named in the *Principles**, some exceed in multitude the number of the sand which is equal in magnitude to the sphere referred to, provided that the following assumptions be made.

1. *The perimeter of the earth is about* 3,000,000 *stadia and not greater.*

It is true that some have tried, as you are of course aware, to prove that the said perimeter is about 300,000 stadia. But I go further and, putting the magnitude of the earth at ten times the size that my predecessors thought it, I suppose its perimeter to be about 3,000,000 stadia and not greater.

---

* 'Αρχαί was apparently the title of the work sent to Zeuxippus. Cf. the note attached to the enumeration of lost works of Archimedes in the Introduction, Chapter II., *ad fin.*

2. *The diameter of the earth is greater than the diameter of the moon, and the diameter of the sun is greater than the diameter of the earth.*

In this assumption I follow most of the earlier astronomers.

3. *The diameter of the sun is about 30 times the diameter of the moon and not greater.*

It is true that, of the earlier astronomers, Eudoxus declared it to be about nine times as great, and Pheidias my father* twelve times, while Aristarchus tried to prove that the diameter of the sun is greater than 18 times but less than 20 times the diameter of the moon. But I go even further than Aristarchus, in order that the truth of my proposition may be established beyond dispute, and I suppose the diameter of the sun to be about 30 times that of the moon and not greater.

4. *The diameter of the sun is greater than the side of the chiliagon inscribed in the greatest circle in the (sphere of the) universe.*

I make this assumption† because Aristarchus discovered that the sun appeared to be about $\frac{1}{720}$th part of the circle of the zodiac, and I myself tried, by a method which I will now describe, to find experimentally (ὀργανικῶς) the angle subtended by the sun and having its vertex at the eye (τὰν γωνίαν, εἰς ἂν ὁ ἅλιος ἐναρμόζει τὰν κορυφὰν ἔχουσαν ποτὶ τᾷ ὄψει)."

[Up to this point the treatise has been literally translated because of the historical interest attaching to the *ipsissima verba* of Archimedes on such a subject. The rest of the work can now be more freely reproduced, and, before proceeding to the mathematical contents of it, it is only necessary to remark that Archimedes next describes how he arrived at a higher and a lower limit for the angle subtended by the sun. This he did

---

* τοῦ ἀμοῦ πατρὸς is the correction of Blass for τοῦ 'Ακούπατρος (*Jahrb. f. Philol.* cxxvii. 1883).

† This is not, strictly speaking, an assumption; it is a proposition proved later (pp. 224—6) by means of the result of an experiment about to be described.

by taking a long rod or ruler (κανών), fastening on the end of it a small cylinder or disc, pointing the rod in the direction of the sun just after its rising (so that it was possible to look directly at it), then putting the cylinder at such a distance that it just concealed, and just failed to conceal, the sun, and lastly measuring the angles subtended by the cylinder. He explains also the correction which he thought it necessary to make because "the eye does not see from one point but from a certain area" (ἐπεὶ αἱ ὄψιες οὐκ ἀφ' ἑνὸς σαμείου βλέποντι, ἀλλὰ ἀπό τινος μεγέθεος).]

The result of the experiment was to show that the angle subtended by the diameter of the sun was less than $\frac{1}{164}$th part, and greater than $\frac{1}{200}$th part, of a right angle.

To prove that (on this assumption) the diameter of the sun is greater than the side of a chiliagon, or figure with 1000 equal sides, inscribed in a great circle of the 'universe.'

Suppose the plane of the paper to be the plane passing through the centre of the sun, the centre of the earth and the eye, at the time when the sun has just risen above the horizon. Let the plane cut the earth in the circle EHL and the sun in the circle FKG, the centres of the earth and sun being C, O respectively, and E being the position of the eye.

Further, let the plane cut the sphere of the 'universe' (i.e. the sphere whose centre is C and radius CO) in the great circle AOB.

Draw from E two tangents to the circle FKG touching it at P, Q, and from C draw two other tangents to the same circle touching it in F, G respectively.

Let CO meet the sections of the earth and sun in H, K respectively; and let CF, CG produced meet the great circle AOB in A, B.

Join EO, OF, OG, OP, OQ, AB, and let AB meet CO in M.

Now $CO > EO$, since the sun is just above the horizon.
Therefore $\angle PEQ > \angle FCG$.

And $\angle PEQ > \frac{1}{200}R$
but $\qquad\qquad < \frac{1}{164}R$ where R represents a right angle.

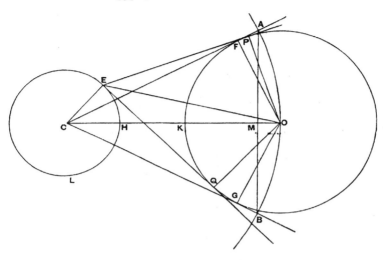

Thus $\qquad\qquad \angle FCG < \frac{1}{164}R$, *a fortiori*,

and the chord AB subtends an arc of the great circle which is less than $\frac{1}{656}$th of the circumference of that circle, i.e.

$AB <$ (side of 656-sided polygon inscribed in the circle).

Now the perimeter of any polygon inscribed in the great circle is less than $\frac{44}{7}CO$. [Cf. *Measurement of a circle*, Prop. 3.]

Therefore $\qquad AB : CO < 11 : 1148$,

and, *a fortiori*, $\qquad AB < \frac{1}{100}CO$ ....(α).

Again, since $CA = CO$, and AM is perpendicular to CO, while OF is perpendicular to CA,

$$AM = OF.$$

Therefore $\quad AB = 2AM =$ (diameter of sun).

Thus $\qquad$ (diameter of sun) $< \frac{1}{100}CO$, by (α),

and, *a fortiori*,

$\qquad\qquad$ (diameter of earth) $< \frac{1}{100}CO$. [Assumption 2]

Hence	$CH + OK < \frac{1}{100}CO,$
so that	$HK > \frac{99}{100}CO,$
or	$CO : HK < 100 : 99.$
And	$CO > CF,$
while	$HK < EQ.$
Therefore	$CF : EQ < 100 : 99$ ....(β).

Now in the right-angled triangles CFO, EQO, of the sides about the right angles,

$$OF = OQ, \text{ but } EQ < CF \text{ (since } EO < CO\text{).}$$

Therefore	$\angle OEQ : \angle OCF > CO : EO,$
but	$< CF : EQ\text{*}.$

Doubling the angles,

$$\angle PEQ : \angle ACB < CF : EQ$$
$$< 100 : 99, \text{ by } (\beta) \text{ above.}$$

But	$\angle PEQ > \frac{1}{200}R,$ by hypothesis.
Therefore	$\angle ACB > \frac{99}{20000}R$
	$> \frac{1}{203}R.$

It follows that the arc AB is greater than $\frac{1}{812}$th of the circumference of the great circle AOB.

Hence, *a fortiori*,

$AB >$ (side of chiliagon inscribed in great circle),

and AB is equal to the diameter of the sun, as proved above.

The following results can now be proved:

(*diameter of 'universe'*) $< 10{,}000$ (*diameter of earth*),

and (*diameter of 'universe'*) $< 10{,}000{,}000{,}000$ *stadia*.

---

* The proposition here assumed is of course equivalent to the trigonometrical formula which states that, if a, β are the circular measures of two angles, each less than a right angle, of which a is the greater, then

$$\frac{\tan a}{\tan \beta} > \frac{a}{\beta} > \frac{\sin a}{\sin \beta}.$$

(1) Suppose, for brevity, that d_u represents the diameter of the 'universe,' d_s that of the sun, d_e that of the earth, and d_m that of the moon.

By hypothesis, $d_s \not> 30 d_m$, [Assumption 3]
and $d_e > d_m$; [Assumption 2]
therefore $d_s < 30 d_e$.

Now, by the last proposition,

$d_s >$ (side of chiliagon inscribed in great circle),
so that (perimeter of chiliagon) $< 1000 d_s$
$< 30{,}000 d_e.$

But the perimeter of any regular polygon with more sides than 6 inscribed in a circle is greater than that of the inscribed regular hexagon, and therefore greater than three times the diameter. Hence

(perimeter of chiliagon) $> 3 d_u$.

It follows that $d_u < 10{,}000 d_e$.

(2) (Perimeter of earth) $\not> 3{,}000{,}000$ stadia.
[Assumption 1]
and (perimeter of earth) $> 3 d_e$.
Therefore $d_e < 1{,}000{,}000$ stadia,
whence $d_u < 10{,}000{,}000{,}000$ stadia.

Assumption 5.

Suppose a quantity of sand taken not greater than a poppy-seed, and suppose that it contains not more than 10,000 grains.

Next suppose the diameter of the poppy-seed to be not less than $\frac{1}{40}$th of a finger-breadth.

Orders and periods of numbers.

I. We have traditional names for numbers up to a myriad (10,000); we can therefore express numbers up to a myriad myriads (100,000,000). Let these numbers be called numbers of the *first order*.

Suppose the 100,000,000 to be the unit of the *second order*, and let the *second order* consist of the numbers from that unit up to $(100{,}000{,}000)^2$.

Let this again be the unit of the *third order* of numbers ending with $(100,000,000)^3$; and so on, until we reach the 100,000,000th *order* of numbers ending with $(100,000,000)^{100,000,000}$, which we will call P.

II. Suppose the numbers from 1 to P just described to form the *first period*.

Let P be the unit of the *first order of the second period*, and let this consist of the numbers from P up to $100,000,000 P$.

Let the last number be the unit of the *second order of the second period*, and let this end with $(100,000,000)^2 P$.

We can go on in this way till we reach the 100,000,000th *order of the second period* ending with $(100,000,000)^{100,000,000} P$, or P^2.

III. Taking P^2 as the unit of the *first order of the third period*, we proceed in the same way till we reach the 100,000,000th *order of the third period* ending with P^3.

IV. Taking P^3 as the unit of the *first order of the fourth period*, we continue the same process until we arrive at the 100,000,000th *order of the* 100,000,000th *period* ending with $P^{100,000,000}$. This last number is expressed by Archimedes as "a myriad-myriad units of the myriad-myriad-th order of the myriad-myriad-th period (αἱ μυριακισμυριοστᾶς περιόδου μυριακισμυριοστῶν ἀριθμῶν μυρίαι μυριάδες)," which is easily seen to be 100,000,000 times the product of $(100,000,000)^{99,999,999}$ and $P^{99,999,999}$, i.e. $P^{100,000,000}$.

[The scheme of numbers thus described can be exhibited more clearly by means of *indices* as follows.

FIRST PERIOD.

First order. Numbers from 1 to 10^8.

Second order. „ „ 10^8 to 10^{16}.

Third order. „ „ 10^{16} to 10^{24}.

⋮

$(10^8)th$ *order.* „ „ $10^{8 \cdot (10^8 - 1)}$ to $10^{8 \cdot 10^8}$ (P, say).

SECOND PERIOD.
   First order.   Numbers from $P.1$ to $P.10^8$.
   Second order.   ,,   ,, $P.10^8$ to $P.10^{16}$.
      ⋮

   $(10^8)th$ *order.*   ,,   ,, $P.10^{8 \cdot (10^8-1)}$ to
                                 $P.10^{8 \cdot 10^8}$ (or P^2).
      ⋮

(10^8)TH PERIOD.
   First order.   ,,   ,, $P^{10^8-1}.1$ to $P^{10^8-1}.10^8$.
   Second order.   ,,   ,, $P^{10^8-1}.10^8$ to $P^{10^8-1}.10^{16}$.
      ⋮

   $(10^8)th$ *order.*   ,,   ,, $P^{10^8-1}.10^{8 \cdot (10^8-1)}$ to
                                   $P^{10^8-1}.10^{8 \cdot 10^8}$ (i.e. P^{10^8}).

The prodigious extent of this scheme will be appreciated when it is considered that the last number in the *first period* would be represented now by 1 followed by 800,000,000 ciphers, while the last number of the $(10^8)th$ *period* would require 100,000,000 times as many ciphers, i.e. 80,000 million millions of ciphers.]

Octads.

Consider the series of terms in continued proportion of which the first is 1 and the second 10 [i.e. the geometrical progression $1, 10^1, 10^2, 10^3, \ldots$]. The *first octad* of these terms [*i.e.* $1, 10^1, 10^2, \ldots 10^7$] fall accordingly under the *first order of the first period* above described, the *second octad* [i.e. $10^8, 10^9, \ldots 10^{15}$] under the *second order of the first period*, the first term of the octad being the unit of the corresponding order in each case. Similarly for the *third octad*, and so on. We can, in the same way, place any number of octads.

Theorem.

If there be any number of terms of a series in continued proportion, say $A_1, A_2, A_3, \ldots A_m, \ldots A_n, \ldots A_{m+n-1}, \ldots$ of which $A_1 = 1$, $A_2 = 10$ [so that the series forms the geometrical progression $1, 10^1, 10^2, \ldots 10^{m-1}, \ldots 10^{n-1}, \ldots 10^{m+n-2}, \ldots$], *and if any two terms as A_m, A_n be taken and multiplied, the product*

$A_m . A_n$ will be a term in the same series and will be as many terms distant from A_n as A_m is distant from A_1; also it will be distant from A_1 by a number of terms less by one than the sum of the numbers of terms by which A_m and A_n respectively are distant from A_1.

Take the term which is distant from A_n by the same number of terms as A_m is distant from A_1. This number of terms is m (the first and last being both counted). Thus the term to be taken is m terms distant from A_n, and is therefore the term A_{m+n-1}.

We have therefore to prove that
$$A_m . A_n = A_{m+n-1}.$$

Now terms equally distant from other terms in the continued proportion are proportional.

Thus $\qquad\qquad \dfrac{A_m}{A_1} = \dfrac{A_{m+n-1}}{A_n}.$

But $\qquad\qquad A_m = A_m . A_1$, since $A_1 = 1$.

Therefore $\qquad A_{m+n-1} = A_m . A_n$ .... (1).

The second result is now obvious, since A_m is m terms distant from A_1, A_n is n terms distant from A_1, and A_{m+n-1} is $(m + n - 1)$ terms distant from A_1.

Application to the number of the sand.

By Assumption 5 [p. 227],

(diam. of poppy-seed) $\not<$ $\frac{1}{40}$ (finger-breadth);

and, since spheres are to one another in the triplicate ratio of their diameters, it follows that

(sphere of diam. 1 finger-breadth) $\not>$ 64,000 poppy-seeds
$\not>$ 64,000 × 10,000
$\not>$ 640,000,000
$\not>$ 6 units of *second order* + 40,000,000 units of *first order*
(*a fortiori*) < 10 units of *second order* of numbers.

⎫
⎬ grains of sand.
⎭

We now gradually increase the diameter of the supposed sphere, multiplying it by 100 each time. Thus, remembering that the sphere is thereby multiplied by 100^3 or 1,000,000, the number of grains of sand which would be contained in a sphere with each successive diameter may be arrived at as follows.

Diameter of sphere.	Corresponding number of grains of sand.
(1) 100 finger-breadths	$<$ 1,000,000 × 10 units of *second order*
	$<$ (7th term of series) × (10th term of series)
	$<$ 16th term of series [i.e. 10^{15}]
	$<$ [10^7 or] 10,000,000 units of the *second order*.
(2) 10,000 finger-breadths	$<$ 1,000,000 × (last number)
	$<$ (7th term of series) × (16th term)
	$<$ 22nd term of series [i.e. 10^{21}]
	$<$ [10^5 or] 100,000 units of *third order*.
(3) 1 stadium ($<$ 10,000 finger-breadths)	$<$ 100,000 units of *third order*.
(4) 100 stadia	$<$ 1,000,000 × (last number)
	$<$ (7th term of series) × (22nd term)
	$<$ 28th term of series [10^{27}]
	$<$ [10^3 or] 1,000 units of *fourth order*.
(5) 10,000 stadia	$<$ 1,000,000 × (last number)
	$<$ (7th term of series) × (28th term)
	$<$ 34th term of series [10^{33}]
	$<$ 10 units of *fifth order*.
(6) 1,000,000 stadia	$<$ (7th term of series) × (34th term)
	$<$ 40th term [10^{39}]
	$<$ [10^7 or] 10,000,000 units of *fifth order*.
(7) 100,000,000 stadia	$<$ (7th term of series) × (40th term)
	$<$ 46th term [10^{45}]
	$<$ [10^5 or] 100,000 units of *sixth order*.
(8) 10,000,000,000 stadia	$<$ (7th term of series) × (46th term)
	$<$ 52nd term of series [10^{51}]
	$<$ [10^3 or] 1,000 units of *seventh order*.

But, by the proposition above [p. 227],

(diameter of 'universe') $<$ 10,000,000,000 stadia.

Hence *the number of grains of sand which could be contained in a sphere of the size of our 'universe' is less than* **1,000** *units of the seventh order of numbers* [or 10^{51}].

From this we can prove further that *a sphere of the size attributed by Aristarchus to the sphere of the fixed stars would contain a number of grains of sand less than* 10,000,000 *units of the eighth order of numbers* [or $10^{56+7} = 10^{63}$].

For, by hypothesis,

(earth) : ('universe') = ('universe') : (sphere of fixed stars).

And [p. 227]

(diameter of 'universe') < 10,000 (diam. of earth);

whence

(diam. of sphere of fixed stars) < 10,000 (diam. of 'universe').

Therefore

(sphere of fixed stars) < $(10,000)^3$. ('universe').

It follows that the number of grains of sand which would be contained in a sphere equal to the sphere of the fixed stars

< $(10,000)^3 \times 1,000$ units of *seventh order*

< (13th term of series) × (52nd term of series)

< 64th term of series [i.e. 10^{63}]

< [10^7 or] 10,000,000 units of *eighth order* of numbers.

Conclusion.

"I conceive that these things, king Gelon, will appear incredible to the great majority of people who have not studied mathematics, but that to those who are conversant therewith and have given thought to the question of the distances and sizes of the earth the sun and moon and the whole universe the proof will carry conviction. And it was for this reason that I thought the subject would be not inappropriate for your consideration."

QUADRATURE OF THE PARABOLA.

"ARCHIMEDES to Dositheus greeting.

"When I heard that Conon, who was my friend in his lifetime, was dead, but that you were acquainted with Conon and withal versed in geometry, while I grieved for the loss not only of a friend but of an admirable mathematician, I set myself the task of communicating to you, as I had intended to send to Conon, a certain geometrical theorem which had not been investigated before but has now been investigated by me, and which I first discovered by means of mechanics and then exhibited by means of geometry. Now some of the earlier geometers tried to prove it possible to find a rectilineal area equal to a given circle and a given segment of a circle; and after that they endeavoured to square the area bounded by the section of the whole cone* and a straight line, assuming lemmas not easily conceded, so that it was recognised by most people that the problem was not solved. But I am not aware that any one of my predecessors has attempted to square the segment bounded by a straight line and a section of a right-angled cone [a parabola], of which problem I have now discovered the solution. For it is here shown that every segment bounded by a straight line and a section of a right-angled cone [a parabola] is four-thirds of the triangle which has the same base and equal height with the segment, and for the demonstration

---

* There appears to be some corruption here: the expression in the text is τᾶς ὅλου τοῦ κώνου τομᾶς, and it is not easy to give a natural and intelligible meaning to it. The section of 'the whole cone' might perhaps mean a section cutting right through it, i.e. an ellipse, and the 'straight line' might be an axis or a diameter. But Heiberg objects to the suggestion to read τᾶς ὀξυγωνίου κώνου τομᾶς, in view of the addition of καὶ εὐθείας, on the ground that the former expression always signifies the whole of an ellipse, never a segment of it (*Quaestiones Archimedeae*, p. 149).

of this property the following lemma is assumed: that the excess by which the greater of (two) unequal areas exceeds the less can, by being added to itself, be made to exceed any given finite area. The earlier geometers have also used this lemma; for it is by the use of this same lemma that they have shown that circles are to one another in the duplicate ratio of their diameters, and that spheres are to one another in the triplicate ratio of their diameters, and further that every pyramid is one third part of the prism which has the same base with the pyramid and equal height; also, that every cone is one third part of the cylinder having the same base as the cone and equal height they proved by assuming a certain lemma similar to that aforesaid. And, in the result, each of the aforesaid theorems has been accepted* no less than those proved without the lemma. As therefore my work now published has satisfied the same test as the propositions referred to, I have written out the proof and send it to you, first as investigated by means of mechanics, and afterwards too as demonstrated by geometry. Prefixed are, also, the elementary propositions in conics which are of service in the proof (στοιχεῖα κωνικὰ χρείαν ἔχοντα ἐς τὰν ἀπόδειξιν). Farewell."

Proposition 1.

If from a point on a parabola a straight line be drawn which is either itself the axis or parallel to the axis, as PV, and if QQ' be a chord parallel to the tangent to the parabola at P and meeting PV in V, then

$$QV = VQ'.$$

Conversely, *if $QV = VQ'$, the chord QQ' will be parallel to the tangent at P.*

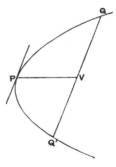

---

* The Greek of this passage is: συμβαίνει δὲ τῶν προειρημένων θεωρημάτων ἕκαστον μηδὲν ἧσσον τῶν ἄνευ τούτου τοῦ λήμματος ἀποδεδειγμένων πεπιστευκέναι. Here it would seem that πεπιστευκέναι must be wrong and that the passive should have been used.

Proposition 2.

If in a parabola QQ' be a chord parallel to the tangent at P, and if a straight line be drawn through P which is either itself the axis or parallel to the axis, and which meets QQ' in V and the tangent at Q to the parabola in T, then

$$PV = PT.$$

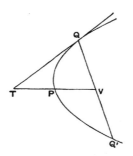

Proposition 3.

If from a point on a parabola a straight line be drawn which is either itself the axis or parallel to the axis, as PV, and if from two other points Q, Q' on the parabola straight lines be drawn parallel to the tangent at P and meeting PV in V, V' respectively, then

$$PV : PV' = QV^2 : Q'V'^2.$$

"And these propositions are proved in the elements of conics.*"

Proposition 4.

If Qq be the base of any segment of a parabola, and P the vertex of the segment, and if the diameter through any other point R meet Qq in O and QP (produced if necessary) in F, then

$$QV : VO = OF : FR.$$

Draw the ordinate RW to PV, meeting QP in K.

---

* i.e. in the treatises on conics by Euclid and Aristaeus.

Then $PV : PW = QV^2 : RW^2$;

whence, by parallels,

$$PQ : PK = PQ^2 : PF^2.$$

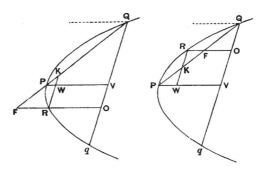

In other words, PQ, PF, PK are in continued proportion; therefore

$$PQ : PF = PF : PK$$
$$= PQ \pm PF : PF \pm PK$$
$$= QF : KF.$$

Hence, by parallels,

$$QV : VO = OF : FR.$$

[It is easily seen that this equation is equivalent to a change of axes of coordinates from the tangent and diameter to new axes consisting of the chord Qq (as axis of x, say) and the diameter through Q (as axis of y).

For, if $QV = a$, $PV = \dfrac{a^2}{p}$, where p is the parameter of the ordinates to PV.

Thus, if $QO = x$, and $RO = y$, the above result gives

$$\frac{a}{x-a} = \frac{OF}{OF-y},$$

whence $$\frac{a}{2a-x} = \frac{OF}{y} = \frac{x \cdot \dfrac{a}{p}}{y},$$

or $$py = x(2a-x).]$$

Proposition 5.

If Qq be the base of any segment of a parabola, P the vertex of the segment, and PV its diameter, and if the diameter of the parabola through any other point R meet Qq in O and the tangent at Q in E, then

$$QO : Oq = ER : RO.$$

Let the diameter through R meet QP in F.

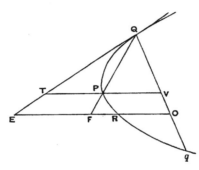

Then, by Prop. 4,
$$QV : VO = OF : FR.$$

Since $QV = Vq$, it follows that
$$QV : qO = OF : OR \quad\ldots\ldots\ldots\ldots\ldots\ldots(1).$$

Also, if VP meet the tangent in T,
$$PT = PV, \text{ and therefore } EF = OF.$$

Accordingly, doubling the antecedents in (1), we have
$$Qq : qO = OE : OR,$$
whence $\qquad QO : Oq = ER : RO.$

Propositions 6, 7*.

Suppose a lever AOB placed horizontally and supported at its middle point O. Let a triangle BCD in which the angle C is right or obtuse be suspended from B and O, so that C is attached to O and CD is in the same vertical line with O. Then, if P be such an area as, when suspended from A, will keep the system in equilibrium,

$$P = \tfrac{1}{3} \triangle BCD.$$

Take a point E on OB such that $BE = 2OE$, and draw EFH parallel to OCD meeting BC, BD in F, H respectively. Let G be the middle point of FH.

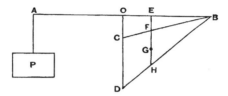

Then G is the centre of gravity of the triangle BCD.

Hence, if the angular points B, C be set free and the triangle be suspended by attaching F to E, the triangle will hang in the same position as before, because EFG is a vertical straight line. "For this is proved†."

Therefore, as before, there will be equilibrium.

Thus $$P : \triangle BCD = OE : AO$$
$$= 1 : 3,$$
or $$P = \tfrac{1}{3} \triangle BCD.$$

---

* In Prop. 6 Archimedes takes the separate case in which the angle BCD of the triangle is a right angle so that C coincides with O in the figure and F with E. He then proves, in Prop. 7, the same property for the triangle in which BCD is an obtuse angle, by treating the triangle as the difference between two right-angled triangles BOD, BOC and using the result of Prop. 6. I have combined the two propositions in one proof, for the sake of brevity. The same remark applies to the propositions following Props. 6, 7.

† Doubtless in the lost book περὶ ζυγῶν. Cf. the Introduction, Chapter II., *ad fin.*

Propositions 8, 9.

Suppose a lever AOB placed horizontally and supported at its middle point O. Let a triangle BCD, right-angled or obtuse-angled at C, be suspended from the points B, E on OB, the angular point C being so attached to E that the side CD is in the same vertical line with E. Let Q be an area such that

$$AO : OE = \triangle BCD : Q.$$

Then, if an area P suspended from A keep the system in equilibrium,

$$P < \triangle BCD \text{ but } > Q.$$

Take G the centre of gravity of the triangle BCD, and draw GH parallel to DC, i.e. vertically, meeting BO in H.

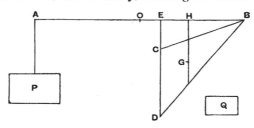

We may now suppose the triangle BCD suspended from H, and, since there is equilibrium,

$$\triangle BCD : P = AO : OH \quad \ldots\ldots\ldots\ldots\ldots\ldots(1),$$

whence $\qquad P < \triangle BCD.$

Also $\qquad \triangle BCD : Q = AO : OE.$

Therefore, by (1), $\triangle BCD : Q > \triangle BCD : P,$

and $\qquad P > Q.$

Propositions 10, 11.

Suppose a lever AOB placed horizontally and supported at O, its middle point. Let CDEF be a trapezium which can be so placed that its parallel sides CD, FE are vertical, while C is vertically below O, and the other sides CF, DE meet in B. Let EF meet BO in H, and let the trapezium be suspended by attaching F to H and C to O. Further, suppose Q to be an area such that

$$AO : OH = (\text{trapezium } CDEF) : Q.$$

Then, if P be the area which, when suspended from A, keeps the system in equilibrium,

$$P < Q.$$

The same is true in the particular case where the angles at C, F are right, and consequently C, F coincide with O, H respectively.

Divide OH in K so that

$$(2CD + FE) : (2FE + CD) = HK : KO.$$

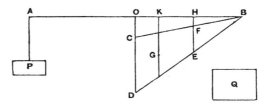

Draw KG parallel to OD, and let G be the middle point of the portion of KG intercepted within the trapezium. Then G is the centre of gravity of the trapezium [*On the equilibrium of planes*, I. 15].

Thus we may suppose the trapezium suspended from K, and the equilibrium will remain undisturbed.

Therefore

$$AO : OK = (\text{trapezium } CDEF) : P,$$

and, by hypothesis,

$$AO : OH = (\text{trapezium } CDEF) : Q.$$

Since $OK < OH$, it follows that

$$P < Q.$$

Propositions 12, 13.

If the trapezium CDEF be placed as in the last propositions, except that CD is vertically below a point L on OB instead of being below O, and the trapezium is suspended from L, H, suppose that Q, R are areas such that

$$AO : OH = (\text{trapezium } CDEF) : Q,$$

and $\quad AO : OL = (\text{trapezium } CDEF) : R.$

If then an area P suspended from A keep the system in equilibrium,

$$P > R \text{ but } < Q.$$

Take the centre of gravity G of the trapezium, as in the last propositions, and let the line through G parallel to DC meet OB in K.

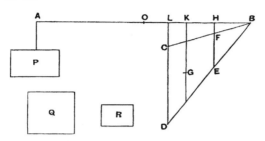

Then we may suppose the trapezium suspended from K, and there will still be equilibrium.

Therefore (trapezium $CDEF$) $: P = AO : OK$.

Hence

(trapezium $CDEF$) $: P >$ (trapezium $CDEF$) $: Q$,

but $\qquad\qquad\qquad\quad <$ (trapezium $CDEF$) $: R$.

It follows that $\qquad P < Q$ but $> R$.

Propositions 14, 15.

Let Qq be the base of any segment of a parabola. Then, if two lines be drawn from Q, q, each parallel to the axis of the parabola and on the same side of Qq as the segment is, either (1) the angles so formed at Q, q are both right angles, or (2) one is acute and the other obtuse. In the latter case let the angle at q be the obtuse angle.

Divide Qq into any number of equal parts at the points $O_1, O_2, \ldots O_n$. Draw through $q, O_1, O_2, \ldots O_n$ diameters of the parabola meeting the tangent at Q in $E, E_1, E_2, \ldots E_n$ and the parabola itself in $q, R_1, R_2, \ldots R_n$. Join $QR_1, QR_2, \ldots QR_n$ meeting $qE, O_1E_1, O_2E_2, \ldots O_{n-1}E_{n-1}$ in $F, F_1, F_2, \ldots F_{n-1}$.

Let the diameters Eq, $E_1O_1, \ldots E_nO_n$ meet a straight line QOA drawn through Q perpendicular to the diameters in the points O, H_1, H_2, $\ldots H_n$ respectively. (In the particular case where Qq is itself perpendicular to the diameters q will coincide with O, O_1 with H_1, and so on.)

It is required to prove that

(1) $\triangle EqQ < 3\,(sum\ of\ trapezia\ FO_1, F_1O_2, \ldots F_{n-1}O_n\ and\ \triangle E_nO_nQ)$,

(2) $\triangle EqQ > 3\,(sum\ of\ trapezia\ R_1O_2, R_2O_3, \ldots R_{n-1}O_n\ and\ \triangle R_nO_nQ)$.

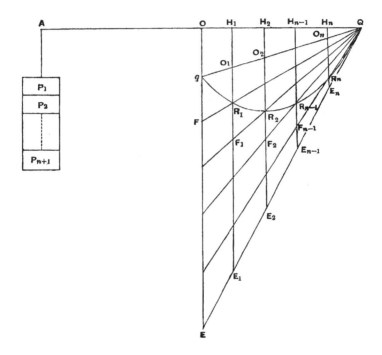

Suppose AO made equal to OQ, and conceive QOA as a lever placed horizontally and supported at O. Suppose the triangle EqQ suspended from OQ in the position drawn, and suppose that the trapezium EO_1 in the position drawn is balanced by an area P_1 suspended from A, the trapezium E_1O_2 in the position drawn is balanced by the area P_2 suspended from A, and so on, the triangle $E_n O_n Q$ being in like manner balanced by P_{n+1}.

Then $P_1 + P_2 + \ldots + P_{n+1}$ will balance the whole triangle EqQ as drawn, and therefore

$$P_1 + P_2 + \ldots + P_{n+1} = \tfrac{1}{3} \triangle EqQ. \quad \text{[Props. 6, 7]}$$

Again $AO : OH_1 = QO : OH_1$
$$= Qq : qO_1$$
$$= E_1 O_1 : O_1 R_1 \text{ [by means of Prop. 5]}$$
$$= \text{(trapezium } EO_1) : \text{(trapezium } FO_1);$$

whence [Props. 10, 11]
$$(FO_1) > P_1.$$

Next $AO : OH_1 = E_1 O_1 : O_1 R_1$
$$= (E_1 O_2) : (R_1 O_2) \ldots \ldots \ldots (\alpha),$$

while $AO : OH_2 = E_2 O_2 : O_2 R_2$
$$= (E_1 O_2) : (F_1 O_2) \ldots \ldots \ldots (\beta);$$

and, since (α) and (β) are simultaneously true, we have, by Props. 12, 13,

$$(F_1 O_2) > P_2 > (R_1 O_2).$$

Similarly it may be proved that

$$(F_2 O_3) > P_3 > (R_2 O_3),$$

and so on.

Lastly [Props. 8, 9]

$$\triangle E_n O_n Q > P_{n+1} > \triangle R_n O_n Q.$$

By addition, we obtain

(1) $(FO_1) + (F_1 O_2) + \ldots + (F_{n-1} O_n) + \triangle E_n O_n Q > P_1 + P_2 + \ldots + P_{n+1}$
$$> \tfrac{1}{3} \triangle EqQ,$$

or $\quad \triangle EqQ < 3 (FO_1 + F_1 O_2 + \ldots + F_{n-1} O_n + \triangle E_n O_n Q).$

(2) $(R_1 O_2) + (R_2 O_3) + \ldots + (R_{n-1} O_n) + \triangle R_n O_n Q < P_2 + P_3 + \ldots + P_{n+1}$
$$< P_1 + P_2 + \ldots + P_{n+1}, \text{ a fortiori,}$$
$$< \tfrac{1}{3} \triangle EqQ,$$

or $\quad \triangle EqQ > 3 (R_1 O_2 + R_2 O_3 + \ldots + R_{n-1} O_n + \triangle R_n O_n Q).$

244 ARCHIMEDES

Proposition 16.

Suppose Qq to be the base of a parabolic segment, q being not more distant than Q from the vertex of the parabola. Draw through q the straight line qE parallel to the axis of the parabola to meet the tangent at Q in E. It is required to prove that

$$(\text{area of segment}) = \tfrac{1}{3} \triangle EqQ.$$

For, if not, the area of the segment must be either greater or less than $\tfrac{1}{3} \triangle EqQ$.

I. Suppose the area of the segment greater than $\tfrac{1}{3} \triangle EqQ$. Then the excess can, if continually added to itself, be made to exceed $\triangle EqQ$. And it is possible to find a submultiple of the triangle EqQ less than the said excess of the segment over $\tfrac{1}{3} \triangle EqQ$.

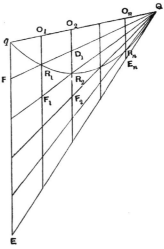

Let the triangle FqQ be such a submultiple of the triangle EqQ. Divide Eq into equal parts each equal to qF, and let all the points of division including F be joined to Q meeting the parabola in $R_1, R_2, \ldots$
R_n respectively. Through $R_1, R_2, \ldots R_n$ draw diameters of the parabola meeting qQ in $O_1, O_2, \ldots O_n$ respectively.

Let O_1R_1 meet QR_2 in F_1.

Let O_2R_2 meet QR_1 in D_1 and QR_3 in F_2.

Let O_3R_3 meet QR_2 in D_2 and QR_4 in F_3, and so on.

We have, by hypothesis,

$$\triangle FqQ < (\text{area of segment}) - \tfrac{1}{3} \triangle EqQ,$$

or $(\text{area of segment}) - \triangle FqQ > \tfrac{1}{3} \triangle EqQ \ldots\ldots\ldots (\alpha)$.

Now, since all the parts of qE, as qF and the rest, are equal, $O_1R_1 = R_1F_1$, $O_2D_1 = D_1R_2 = R_2F_2$, and so on; therefore
$$\triangle FqQ = (FO_1 + R_1O_2 + D_1O_3 + \ldots)$$
$$= (FO_1 + F_1D_1 + F_2D_2 + \ldots + F_{n-1}D_{n-1} + \triangle E_nR_nQ)\ldots(\beta).$$
But

(area of segment) $< (FO_1 + F_1O_2 + \ldots + F_{n-1}O_n + \triangle E_nO_nQ)$.

Subtracting, we have

(area of segment) $- \triangle FqQ < (R_1O_2 + R_2O_3 + \ldots$
$$+ R_{n-1}O_n + \triangle R_nO_nQ),$$
whence, *a fortiori*, by (α),
$$\tfrac{1}{3}\triangle EqQ < (R_1O_2 + R_2O_3 + \ldots + R_{n-1}O_n + \triangle R_nO_nQ).$$
But this is impossible, since [Props. 14, 15]
$$\tfrac{1}{3}\triangle EqQ > (R_1O_2 + R_2O_3 + \ldots + R_{n-1}O_n + \triangle R_nO_nQ).$$
Therefore
$$\text{(area of segment)} \not> \tfrac{1}{3}\triangle EqQ.$$

II. If possible, suppose the area of the segment less than $\tfrac{1}{3}\triangle EqQ$.

Take a submultiple of the triangle EqQ, as the triangle FqQ, less than the excess of $\tfrac{1}{3}\triangle EqQ$ over the area of the segment, and make the same construction as before.

Since $\triangle FqQ < \tfrac{1}{3}\triangle EqQ - \text{(area of segment)}$,
it follows that
$$\triangle FqQ + \text{(area of segment)} < \tfrac{1}{3}\triangle EqQ$$
$$< (FO_1 + F_1O_2 + \ldots + F_{n-1}O_n + \triangle E_nO_nQ).$$
[Props. 14, 15]

Subtracting from each side the area of the segment, we have

$\triangle FqQ < $ (sum of spaces qFR_1, $R_1F_1R_2$, ... E_nR_nQ)
$$< (FO_1 + F_1D_1 + \ldots + F_{n-1}D_{n-1} + \triangle E_nR_nQ), \textit{a fortiori};$$
which is impossible, because, by (β) above,
$$\triangle FqQ = FO_1 + F_1D_1 + \ldots + F_{n-1}D_{n-1} + \triangle E_nR_nQ.$$
Hence (area of segment) $\not< \tfrac{1}{3}\triangle EqQ$.

Since then the area of the segment is neither less nor greater than $\tfrac{1}{3}\triangle EqQ$, it is equal to it.

Proposition 17.

It is now manifest that *the area of any segment of a parabola is four-thirds of the triangle which has the same base as the segment and equal height.*

Let Qq be the base of the segment, P its vertex. Then PQq is the inscribed triangle with the same base as the segment and equal height.

Since P is the vertex* of the segment, the diameter through P bisects Qq. Let V be the point of bisection.

Let VP, and qE drawn parallel to it, meet the tangent at Q in T, E respectively.

Then, by parallels,

$$qE = 2VT,$$
and $$PV = PT, \quad [\text{Prop. 2}]$$
so that $$VT = 2PV.$$

Hence $\triangle EqQ = 4 \triangle PQq$.

But, by Prop. 16, the area of the segment is equal to $\frac{1}{3} \triangle EqQ$.

Therefore (area of segment) $= \frac{4}{3} \triangle PQq$.

DEF. "In segments bounded by a straight line and any curve I call the straight line the **base**, and the **height** the greatest perpendicular drawn from the curve to the base of the segment, and the **vertex** the point from which the greatest perpendicular is drawn."

---

* It is curious that Archimedes uses the terms *base* and *vertex* of a segment here, but gives the definition of them later (at the end of the proposition). Moreover he assumes the converse of the property proved in Prop. 18.

Proposition 18.

If Qq be the base of a segment of a parabola, and V the middle point of Qq, and if the diameter through V meet the curve in P, then P is the vertex of the segment.

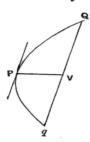

For Qq is parallel to the tangent at P [Prop. 1]. Therefore, of all the perpendiculars which can be drawn from points on the segment to the base Qq, that from P is the greatest. Hence, by the definition, P is the vertex of the segment.

Proposition 19.

If Qq be a chord of a parabola bisected in V by the diameter PV, and if RM be a diameter bisecting QV in M, and RW be the ordinate from R to PV, then
$$PV = \tfrac{4}{3}RM.$$

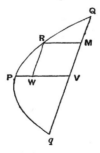

For, by the property of the parabola,
$$PV : PW = QV^2 : RW^2$$
$$= 4RW^2 : RW^2,$$
so that $\qquad PV = 4PW,$
whence $\qquad PV = \tfrac{4}{3}RM.$

Proposition 20.

If Qq be the base, and P the vertex, of a parabolic segment, then the triangle PQq is greater than half the segment PQq.

For the chord Qq is parallel to the tangent at P, and the triangle PQq is half the parallelogram formed by Qq, the tangent at P, and the diameters through Q, q.

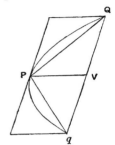

Therefore the triangle PQq is greater than half the segment.

COR. It follows that *it is possible to inscribe in the segment a polygon such that the segments left over are together less than any assigned area.*

Proposition 21.

If Qq be the base, and P the vertex, of any parabolic segment, and if R be the vertex of the segment cut off by PQ, then
$$\triangle PQq = 8\triangle PRQ.$$

The diameter through R will bisect the chord PQ, and therefore also QV, where PV is the diameter bisecting Qq. Let the diameter through R bisect PQ in Y and QV in M. Join PM.

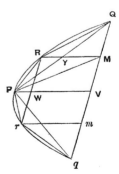

By Prop. 19,
$$PV = \tfrac{4}{3}RM.$$
Also $\qquad PV = 2YM.$
Therefore $\qquad YM = 2RY,$
and $\qquad \triangle PQM = 2\triangle PRQ.$
Hence $\qquad \triangle PQV = 4\triangle PRQ,$
and $\qquad \triangle PQq = 8\triangle PRQ.$

Also, if RW, the ordinate from R to PV, be produced to meet the curve again in r,
$$RW = rW,$$
and the same proof shows that
$$\triangle PQq = 8 \triangle Prq.$$

Proposition 22.

If there be a series of areas $A, B, C, D, \ldots$ each of which is four times the next in order, and if the largest, A, be equal to the triangle PQq inscribed in a parabolic segment PQq and having the same base with it and equal height, then

$$(A + B + C + D + \ldots) < (\textit{area of segment } PQq).$$

For, since $\triangle PQq = 8 \triangle PRQ = 8 \triangle Pqr$, where R, r are the vertices of the segments cut off by PQ, Pq, as in the last proposition,

$$\triangle PQq = 4 (\triangle PQR + \triangle Pqr).$$

Therefore, since $\triangle PQq = A$,

$$\triangle PQR + \triangle Pqr = B.$$

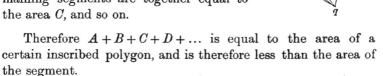

In like manner we prove that the triangles similarly inscribed in the remaining segments are together equal to the area C, and so on.

Therefore $A + B + C + D + \ldots$ is equal to the area of a certain inscribed polygon, and is therefore less than the area of the segment.

Proposition 23.

Given a series of areas $A, B, C, D, \ldots Z$, of which A is the greatest, and each is equal to four times the next in order, then

$$A + B + C + \ldots + Z + \tfrac{1}{3}Z = \tfrac{4}{3}A.$$

Take areas $b, c, d, \ldots$ such that
$$b = \tfrac{1}{3}B,$$
$$c = \tfrac{1}{3}C,$$
$$d = \tfrac{1}{3}D, \text{ and so on.}$$

Then, since $b = \tfrac{1}{3}B,$

and $B = \tfrac{1}{4}A,$

$B + b = \tfrac{1}{3}A.$

Similarly $C + c = \tfrac{1}{3}B.$

....

Therefore

$B + C + D + \ldots + Z + b + c + d + \ldots + z = \tfrac{1}{3}(A + B + C + \ldots + Y).$

But $b + c + d + \ldots + y = \tfrac{1}{3}(B + C + D + \ldots + Y).$

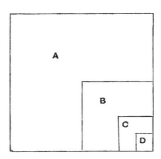

Therefore, by subtraction,

$$B + C + D + \ldots + Z + z = \tfrac{1}{3}A$$

or $A + B + C + \ldots + Z + \tfrac{1}{3}Z = \tfrac{4}{3}A.$

QUADRATURE OF THE PARABOLA. 251

[The algebraical equivalent of this result is of course
$$1 + \tfrac{1}{4} + (\tfrac{1}{4})^2 + \ldots + (\tfrac{1}{4})^{n-1} = \tfrac{4}{3} - \tfrac{1}{3}(\tfrac{1}{4})^{n-1}$$
$$= \frac{1 - (\tfrac{1}{4})^n}{1 - \tfrac{1}{4}}.]$$

Proposition 24.

Every segment bounded by a parabola and a chord Qq is equal to four-thirds of the triangle which has the same base as the segment and equal height.

Suppose $K = \tfrac{4}{3} \triangle PQq$,

where P is the vertex of the segment; and we have then to prove that the area of the segment is equal to K.

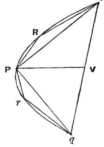

For, if the segment be not equal to K, it must either be greater or less.

I. Suppose the area of the segment greater than K.

If then we inscribe in the segments cut off by PQ, Pq triangles which have the same base and equal height, i.e. triangles with the same vertices R, r as those of the segments, and if in the remaining segments we inscribe triangles in the same manner, and so on, we shall finally have segments remaining whose sum is less than the area by which the segment PQq exceeds K.

Therefore the polygon so formed must be greater than the area K; which is impossible, since [Prop. 23]
$$A + B + C + \ldots + Z < \tfrac{4}{3} A,$$
where $A = \triangle PQq$.

Thus the area of the segment cannot be greater than K.

II. Suppose, if possible, that the area of the segment is less than K.

If then $\triangle PQq = A$, $B = \frac{1}{4}A$, $C = \frac{1}{4}B$, and so on, until we arrive at an area X such that X is less than the difference between K and the segment, we have

$$A + B + C + \ldots + X + \tfrac{1}{3}X = \tfrac{4}{3}A \qquad \text{[Prop. 23]}$$
$$= K.$$

Now, since K exceeds $A + B + C + \ldots + X$ by an area less than X, and the area of the segment by an area greater than X, it follows that

$$A + B + C + \ldots + X > \text{(the segment)};$$

which is impossible, by Prop. 22 above.

Hence the segment is not less than K.

Thus, since the segment is neither greater nor less than K,

(area of segment PQq) $= K = \tfrac{4}{3} \triangle PQq$.

ON FLOATING BODIES.

BOOK I.

Postulate 1.

"Let it be supposed that a fluid is of such a character that, its parts lying evenly and being continuous, that part which is thrust the less is driven along by that which is thrust the more; and that each of its parts is thrust by the fluid which is above it in a perpendicular direction if the fluid be sunk in anything and compressed by anything else."

Proposition 1.

If a surface be cut by a plane always passing through a certain point, and if the section be always a circumference [of a circle] whose centre is the aforesaid point, the surface is that of a sphere.

For, if not, there will be some two lines drawn from the point to the surface which are not equal.

Suppose O to be the fixed point, and A, B to be two points on the surface such that OA, OB are unequal. Let the surface be cut by a plane passing through OA, OB. Then the section is, by hypothesis, a circle whose centre is O.

Thus $OA = OB$; which is contrary to the assumption. Therefore the surface cannot but be a sphere.

Proposition 2.

The surface of any fluid at rest is the surface of a sphere whose centre is the same as that of the earth.

Suppose the surface of the fluid cut by a plane through O, the centre of the earth, in the curve $ABCD$.

$ABCD$ shall be the circumference of a circle.

For, if not, some of the lines drawn from O to the curve will be unequal. Take one of them, OB, such that OB is greater than some of the lines from O to the curve and less than others. Draw a circle with OB as radius. Let it be EBF, which will therefore fall partly within and partly without the surface of the fluid.

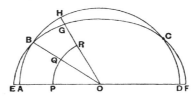

Draw OGH making with OB an angle equal to the angle EOB, and meeting the surface in H and the circle in G. Draw also in the plane an arc of a circle PQR with centre O and within the fluid.

Then the parts of the fluid along PQR are uniform and continuous, and the part PQ is compressed by the part between it and AB, while the part QR is compressed by the part between QR and BH. Therefore the parts along PQ, QR will be unequally compressed, and the part which is compressed the less will be set in motion by that which is compressed the more.

Therefore there will not be rest; which is contrary to the hypothesis.

Hence the section of the surface will be the circumference of a circle whose centre is O; and so will all other sections by planes through O.

Therefore the surface is that of a sphere with centre O.

Proposition 3.

Of solids those which, size for size, are of equal weight with a fluid will, if let down into the fluid, be immersed so that they do not project above the surface but do not sink lower.

If possible, let a certain solid $EFHG$ of equal weight, volume for volume, with the fluid remain immersed in it so that part of it, $EBCF$, projects above the surface.

Draw through O, the centre of the earth, and through the solid a plane cutting the surface of the fluid in the circle $ABCD$.

Conceive a pyramid with vertex O and base a parallelogram at the surface of the fluid, such that it includes the immersed portion of the solid. Let this pyramid be cut by the plane of $ABCD$ in OL, OM. Also let a sphere within the fluid and below GH be described with centre O, and let the plane of $ABCD$ cut this sphere in PQR.

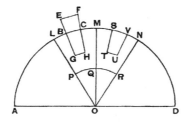

Conceive also another pyramid in the fluid with vertex O, continuous with the former pyramid and equal and similar to it. Let the pyramid so described be cut in OM, ON by the plane of $ABCD$.

Lastly, let $STUV$ be a part of the fluid within the second pyramid equal and similar to the part $BGHC$ of the solid, and let SV be at the surface of the fluid.

Then the pressures on PQ, QR are unequal, that on PQ being the greater. Hence the part at QR will be set in motion by that at PQ, and the fluid will not be at rest; which is contrary to the hypothesis.

Therefore the solid will not stand out above the surface.

Nor will it sink further, because all the parts of the fluid will be under the same pressure.

Proposition 4.

A solid lighter than a fluid will, if immersed in it, not be completely submerged, but part of it will project above the surface.

In this case, after the manner of the previous proposition, we assume the solid, if possible, to be completely submerged and the fluid to be at rest in that position, and we conceive (1) a pyramid with its vertex at O, the centre of the earth, including the solid, (2) another pyramid continuous with the former and equal and similar to it, with the same vertex O, (3) a portion of the fluid within this latter pyramid equal to the immersed solid in the other pyramid, (4) a sphere with centre O whose surface is below the immersed solid and the part of the fluid in the second pyramid corresponding thereto. We suppose a plane to be drawn through the centre O cutting the surface of the fluid in the circle ABC, the solid in S, the first pyramid in OA, OB, the second pyramid in OB, OC, the portion of the fluid in the second pyramid in K, and the inner sphere in PQR.

Then the pressures on the parts of the fluid at PQ, QR are unequal, since S is lighter than K. Hence there will not be rest; which is contrary to the hypothesis.

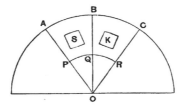

Therefore the solid S cannot, in a condition of rest, be completely submerged.

Proposition 5.

Any solid lighter than a fluid will, if placed in the fluid, be so far immersed that the weight of the solid will be equal to the weight of the fluid displaced.

For let the solid be *EGHF*, and let *BGHC* be the portion of it immersed when the fluid is at rest. As in Prop. 3, conceive a pyramid with vertex *O* including the solid, and another pyramid with the same vertex continuous with the former and equal and similar to it. Suppose a portion of the fluid *STUV* at the base of the second pyramid to be equal and similar to the immersed portion of the solid; and let the construction be the same as in Prop. 3.

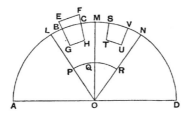

Then, since the pressure on the parts of the fluid at *PQ*, *QR* must be equal in order that the fluid may be at rest, it follows that the weight of the portion *STUV* of the fluid must be equal to the weight of the solid *EGHF*. And the former is equal to the weight of the fluid displaced by the immersed portion of the solid *BGHC*.

Proposition 6.

If a solid lighter than a fluid be forcibly immersed in it, the solid will be driven upwards by a force equal to the difference between its weight and the weight of the fluid displaced.

For let *A* be completely immersed in the fluid, and let *G* represent the weight of *A*, and $(G + H)$ the weight of an equal volume of the fluid. Take a solid *D*, whose weight is *H* and add it to A. Then the weight of $(A + D)$ is less than that of an equal volume of the fluid; and, if $(A + D)$ is immersed in the fluid, it will project so that its weight will be equal to the weight of the fluid displaced. But its weight is $(G + H)$.

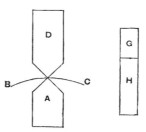

Therefore the weight of the fluid displaced is $(G + H)$, and hence the volume of the fluid displaced is the volume of the solid A. There will accordingly be rest with A immersed and D projecting.

Thus the weight of D balances the upward force exerted by the fluid on A, and therefore the latter force is equal to H, which is the difference between the weight of A and the weight of the fluid which A displaces.

Proposition 7.

A solid heavier than a fluid will, if placed in it, descend to the bottom of the fluid, and the solid will, when weighed in the fluid, be lighter than its true weight by the weight of the fluid displaced.

(1) The first part of the proposition is obvious, since the part of the fluid under the solid will be under greater pressure, and therefore the other parts will give way until the solid reaches the bottom.

(2) Let A be a solid heavier than the same volume of the fluid, and let $(G + H)$ represent its weight, while G represents the weight of the same volume of the fluid.

Take a solid B lighter than the same volume of the fluid, and such that the weight of B is G, while the weight of the same volume of the fluid is $(G+H)$.

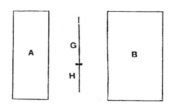

Let A and B be now combined into one solid and immersed. Then, since $(A+B)$ will be of the same weight as the same volume of fluid, both weights being equal to $(G+H)+G$, it follows that $(A+B)$ will remain stationary in the fluid.

Therefore the force which causes A by itself to sink must be equal to the upward force exerted by the fluid on B by itself. This latter is equal to the difference between $(G+H)$ and G [Prop. 6]. Hence A is depressed by a force equal to H, i.e. its weight in the fluid is H, or the difference between $(G+H)$ and G.

[This proposition may, I think, safely be regarded as decisive of the question how Archimedes determined the proportions of gold and silver contained in the famous crown (cf. Introduction, Chapter I.). The proposition suggests in fact the following method.

Let W represent the weight of the crown, w_1 and w_2 the weights of the gold and silver in it respectively, so that $W = w_1 + w_2$.

(1) Take a weight W of pure gold and weigh it in a fluid. The apparent loss of weight is then equal to the weight of the fluid displaced. If F_1 denote this weight, F_1 is thus known as the result of the operation of weighing.

It follows that the weight of fluid displaced by a weight w_1 of gold is $\frac{w_1}{W} \cdot F_1$.

(2) Take a weight W of pure silver and perform the same operation. If F_2 be the loss of weight when the silver is weighed in the fluid, we find in like manner that the weight of fluid displaced by w_2 is $\dfrac{w_2}{W} \cdot F_2$.

(3) Lastly, weigh the crown itself in the fluid, and let F be the loss of weight. Therefore the weight of fluid displaced by the crown is F.

It follows that
$$\dfrac{w_1}{W} \cdot F_1 + \dfrac{w_2}{W} \cdot F_2 = F,$$

or
$$w_1 F_1 + w_2 F_2 = (w_1 + w_2) F,$$

whence
$$\dfrac{w_1}{w_2} = \dfrac{F_2 - F}{F - F_1}.$$

This procedure corresponds pretty closely to that described in the poem *de ponderibus et mensuris* (written probably about 500 A.D.)* purporting to explain Archimedes' method. According to the author of this poem, we first take two equal weights of pure gold and pure silver respectively and weigh them against each other when both immersed in water; this gives the relation between their weights in water and therefore between their loss of weight in water. Next we take the mixture of gold and silver and an equal weight of pure silver and weigh them against each other in water in the same manner.

The other version of the method used by Archimedes is that given by Vitruvius†, according to which he measured successively the *volumes* of fluid displaced by three equal weights, (1) the crown, (2) the same weight of gold, (3) the same weight of silver, respectively. Thus, if as before the weight of the crown is W, and it contains weights w_1 and w_2 of gold and silver respectively,

(1) the crown displaces a certain quantity of fluid, V say.

(2) the weight W of gold displaces a certain volume of

---

* Torelli's *Archimedes*, p. 364; Hultsch, *Metrol. Script.* II. 95 sq., and Prolegomena § 118.

† *De architect.* IX. 3.

fluid, V_1 say; therefore a weight w_1 of gold displaces a volume $\frac{w_1}{W}.V_1$ of fluid.

(3) the weight W of silver displaces a certain volume of fluid, say V_2; therefore a weight w_2 of silver displaces a volume $\frac{w_2}{W}.V_2$ of fluid.

It follows that $\quad V = \frac{w_1}{W}.V_1 + \frac{w_2}{W}.V_2,$

whence, since $\quad W = w_1 + w_2,$

$$\frac{w_1}{w_2} = \frac{V_2 - V}{V - V_1};$$

and this ratio is obviously equal to that before obtained, viz. $\frac{F_2 - F}{F - F_1}$.]

Postulate 2.

"Let it be granted that bodies which are forced upwards in a fluid are forced upwards along the perpendicular [to the surface] which passes through their centre of gravity."

Proposition 8.

If a solid in the form of a segment of a sphere, and of a substance lighter than a fluid, be immersed in it so that its base does not touch the surface, the solid will rest in such a position that its axis is perpendicular to the surface; and, if the solid be forced into such a position that its base touches the fluid on one side and be then set free, it will not remain in that position but will return to the symmetrical position.

[The proof of this proposition is wanting in the Latin version of Tartaglia. Commandinus supplied a proof of his own in his edition.]

Proposition 9.

If a solid in the form of a segment of a sphere, and of a substance lighter than a fluid, be immersed in it so that its base is completely below the surface, the solid will rest in such a position that its axis is perpendicular to the surface.

[The proof of this proposition has only survived in a mutilated form. It deals moreover with only one case out of three which are distinguished at the beginning, viz. that in which the segment is greater than a hemisphere, while figures only are given for the cases where the segment is equal to, or less than, a hemisphere.]

Suppose, first, that the segment is greater than a hemisphere. Let it be cut by a plane through its axis and the centre of the earth; and, if possible, let it be at rest in the position shown in the figure, where AB is the intersection of the plane with the base of the segment, DE its axis, C the centre of the sphere of which the segment is a part, O the centre of the earth.

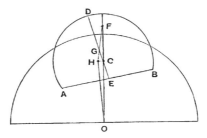

The centre of gravity of the portion of the segment outside the fluid, as F, lies on OC produced, its axis passing through C.

Let G be the centre of gravity of the segment. Join FG, and produce it to H so that

$FG : GH =$ (volume of immersed portion) : (rest of solid).

Join OH.

Then the weight of the portion of the solid outside the fluid acts along FO, and the pressure of the fluid on the immersed portion along OH, while the weight of the immersed portion acts along HO and is by hypothesis less than the pressure of the fluid acting along OH.

Hence there will not be equilibrium, but the part of the segment towards A will ascend and the part towards B descend, until DE assumes a position perpendicular to the surface of the fluid.

ON FLOATING BODIES.

BOOK II.

Proposition 1.

If a solid lighter than a fluid be at rest in it, the weight of the solid will be to that of the same volume of the fluid as the immersed portion of the solid is to the whole.

Let $(A + B)$ be the solid, B the portion immersed in the fluid.

Let $(C + D)$ be an equal volume of the fluid, C being equal in volume to A and B to D.

Further suppose the line E to represent the weight of the solid $(A + B)$, $(F + G)$ to represent the weight of $(C + D)$, and G that of D.

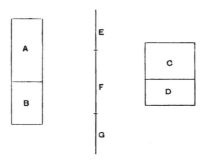

Then

weight of $(A + B)$: weight of $(C + D) = E : (F + G)$...(1).

And the weight of $(A + B)$ is equal to the weight of a volume B of the fluid [I. 5], i.e. to the weight of D.

That is to say, $E = G$.

Hence, by (1),

weight of $(A + B)$: weight of $(C + D) = G : F + G$
$$= D : C + D$$
$$= B : A + B.$$

Proposition 2.

If a right segment of a paraboloid of revolution whose axis is not greater than $\frac{3}{4} p$ (where p is the principal parameter of the generating parabola), and whose specific gravity is less than that of a fluid, be placed in the fluid with its axis inclined to the vertical at any angle, but so that the base of the segment does not touch the surface of the fluid, the segment of the paraboloid will not remain in that position but will return to the position in which its axis is vertical.

Let the axis of the segment of the paraboloid be AN, and through AN draw a plane perpendicular to the surface of the fluid. Let the plane intersect the paraboloid in the parabola BAB', the base of the segment of the paraboloid in BB', and the plane of the surface of the fluid in the chord QQ' of the parabola.

Then, since the axis AN is placed in a position not perpendicular to QQ', BB' will not be parallel to QQ'.

Draw the tangent PT to the parabola which is parallel to QQ', and let P be the point of contact*.

[From P draw PV parallel to AN meeting QQ' in V. Then PV will be a diameter of the parabola, and also the axis of the portion of the paraboloid immersed in the fluid.

---

* The rest of the proof is wanting in the version of Tartaglia, but is given in brackets as supplied by Commandinus.

Let C be the centre of gravity of the paraboloid BAB', and F that of the portion immersed in the fluid. Join FC and produce it to H so that H is the centre of gravity of the remaining portion of the paraboloid above the surface.

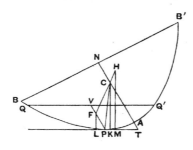

Then, since $AN = \tfrac{3}{2}AC^*$,

and $AN \not> \tfrac{3}{4}p$,

it follows that $AC \not> \dfrac{p}{2}$.

Therefore, if CP be joined, the angle CPT is acute†. Hence, if CK be drawn perpendicular to PT, K will fall between P and T. And, if FL, HM be drawn parallel to CK to meet PT, they will each be perpendicular to the surface of the fluid.

Now the force acting on the immersed portion of the segment of the paraboloid will act upwards along LF, while the weight of the portion outside the fluid will act downwards along HM.

Therefore there will not be equilibrium, but the segment

* As the determination of the centre of gravity of a segment of a paraboloid which is here assumed does not appear in any extant work of Archimedes, or in any known work by any other Greek mathematician, it appears probable that it was investigated by Archimedes himself in some treatise now lost.

† The truth of this statement is easily proved from the property of the subnormal. For, if the normal at P meet the axis in G, AG is greater than $\dfrac{p}{2}$ except in the case where the normal is the normal at the vertex A itself. But the latter case is excluded here because, by hypothesis, AN is not placed vertically. Hence, P being a different point from A, AG is always greater than AC; and, since the angle TPG is right, the angle TPC must be acute.

will turn so that B will rise and B' will fall, until AN takes the vertical position.]

[For purposes of comparison the trigonometrical equivalent of this and other propositions will be appended.

Suppose that the angle NTP, at which in the above figure the axis AN is inclined to the surface of the fluid, is denoted by θ.

Then the coordinates of P referred to AN and the tangent at A as axes are

$$\frac{p}{4}\cot^2\theta,\quad \frac{p}{2}\cot\theta,$$

where p is the principal parameter.

Suppose that $AN = h$, $PV = k$.

If now x' be the distance from T of the orthogonal projection of F on TP, and x the corresponding distance for the point C, we have

$$x' = \frac{p}{2}\cot^2\theta \cdot \cos\theta + \frac{p}{2}\cot\theta \cdot \sin\theta + \frac{2}{3}k\cos\theta,$$

$$x = \frac{p}{4}\cot^2\theta \cdot \cos\theta + \frac{2}{3}h\cos\theta,$$

whence $\quad x' - x = \cos\theta\left\{\frac{p}{4}(\cot^2\theta + 2) - \frac{2}{3}(h-k)\right\}.$

In order that the segment of the paraboloid may turn in the direction of increasing the angle PTN, x' must be greater than x, or the expression just found must be positive.

This will always be the case, whatever be the value of θ, if

$$\frac{p}{2} \not< \frac{2h}{3},$$

or $\qquad h \not> \tfrac{3}{4}p.$]

Proposition 3.

If a right segment of a paraboloid of revolution whose axis is not greater than $\tfrac{3}{4}p$ (where p is the parameter), and whose specific gravity is less than that of a fluid, be placed in the fluid with its axis inclined at any angle to the vertical, but so that its *base is entirely submerged, the solid will not remain in that position but will return to the position in which the axis is vertical.*

Let the axis of the paraboloid be AN, and through AN draw a plane perpendicular to the surface of the fluid intersecting the paraboloid in the parabola BAB', the base of the segment in BNB', and the plane of the surface of the fluid in the chord QQ' of the parabola.

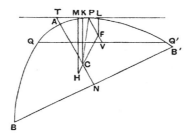

Then, since AN, as placed, is not perpendicular to the surface of the fluid, QQ' and BB' will not be parallel.

Draw PT parallel to QQ' and touching the parabola at P. Let PT meet NA produced in T. Draw the diameter PV bisecting QQ' in V. PV is then the axis of the portion of the paraboloid above the surface of the fluid.

Let C be the centre of gravity of the whole segment of the paraboloid, F that of the portion above the surface. Join FC and produce it to H so that H is the centre of gravity of the immersed portion.

Then, since $AC \not> \frac{p}{2}$, the angle CPT is an acute angle, as in the last proposition.

Hence, if CK be drawn perpendicular to PT, K will fall between P and T. Also, if HM, FL be drawn parallel to CK, they will be perpendicular to the surface of the fluid.

And the force acting on the submerged portion will act upwards along HM, while the weight of the rest will act downwards along LF produced.

Thus the paraboloid will turn until it takes the position in which AN is vertical.

Proposition 4.

Given a right segment of a paraboloid of revolution whose axis AN is greater than $\frac{3}{4}p$ (where p is the parameter), and whose specific gravity is less than that of a fluid but bears to it a ratio not less than $(AN - \frac{3}{4}p)^2 : AN^2$, if the segment of the paraboloid be placed in the fluid with its axis at any inclination to the vertical, but so that its base does not touch the surface of the fluid, it will not remain in that position but will return to the position in which its axis is vertical.

Let the axis of the segment of the paraboloid be AN, and let a plane be drawn through AN perpendicular to the surface of the fluid and intersecting the segment in the parabola BAB', the base of the segment in BB', and the surface of the fluid in the chord QQ' of the parabola.

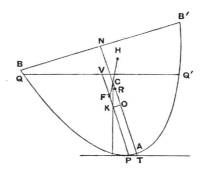

Then AN, as placed, will not be perpendicular to QQ'.

Draw PT parallel to QQ' and touching the parabola at P. Draw the diameter PV bisecting QQ' in V. Thus PV will be the axis of the submerged portion of the solid.

Let C be the centre of gravity of the whole solid, F that of the immersed portion. Join FC and produce it to H so that H is the centre of gravity of the remaining portion.

Now, since $\qquad AN = \frac{3}{2}AC$,

and $\qquad AN > \frac{3}{4}p$,

it follows that $\qquad AC > \dfrac{p}{2}$.

Measure CO along CA equal to $\frac{p}{2}$, and OR along OC equal to $\frac{1}{2}AO$.

Then, since $\quad AN = \frac{3}{2}AC,$

and $\quad AR = \frac{3}{2}AO,$

we have, by subtraction,
$$NR = \frac{3}{2}OC.$$

That is, $\quad AN - AR = \frac{3}{2}OC$
$$= \frac{3}{4}p,$$

or $\quad AR = (AN - \frac{3}{4}p).$

Thus $\quad (AN - \frac{3}{4}p)^2 : AN^2 = AR^2 : AN^2,$

and therefore the ratio of the specific gravity of the solid to that of the fluid is, by the enunciation, not less than the ratio $AR^2 : AN^2$.

But, by Prop. 1, the former ratio is equal to the ratio of the immersed portion to the whole solid, i.e. to the ratio $PV^2 : AN^2$ [*On Conoids and Spheroids*, Prop. 24].

Hence $\quad PV^2 : AN^2 \not< AR^2 : AN^2,$

or $\quad PV \not< AR.$

It follows that
$$PF(=\tfrac{2}{3}PV) \not< \tfrac{2}{3}AR$$
$$\not< AO.$$

If, therefore, OK be drawn from O perpendicular to OA, it will meet PF between P and F.

Also, if CK be joined, the triangle KCO is equal and similar to the triangle formed by the normal, the subnormal and the ordinate at P (since $CO = \frac{1}{2}p$ or the subnormal, and KO is equal to the ordinate).

Therefore CK is parallel to the normal at P, and therefore perpendicular to the tangent at P and to the surface of the fluid.

Hence, if parallels to CK be drawn through F, H, they will be perpendicular to the surface of the fluid, and the force acting on the submerged portion of the solid will act upwards along the former, while the weight of the other portion will act downwards along the latter.

Therefore the solid will not remain in its position but will turn until AN assumes a vertical position.

[Using the same notation as before (note following Prop. 2), we have
$$x' - x = \cos\theta \left\{ \frac{p}{4}(\cot^2\theta + 2) - \frac{2}{3}(h-k) \right\},$$
and the *minimum* value of the expression within the bracket, for different values of θ, is
$$\frac{p}{2} - \frac{2}{3}(h-k),$$
corresponding to the position in which AM is vertical, or $\theta = \frac{\pi}{2}$. Therefore there will be stable equilibrium in that position only, provided that
$$k \not< (h - \tfrac{3}{4}p),$$
or, if s be the ratio of the specific gravity of the solid to that of the fluid ($= k^2/h^2$ in this case),
$$s \not< (h - \tfrac{3}{4}p)^2/h^2.]$$

Proposition 5.

Given a right segment of a paraboloid of revolution such that its axis AN is greater than $\tfrac{3}{4}p$ (where p is the parameter), and its specific gravity is less than that of a fluid but in a ratio to it not greater than the ratio $\{AN^2 - (AN - \tfrac{3}{4}p)^2\} : AN^2$, if the segment be placed in the fluid with its axis inclined at any angle to the vertical, but so that its base is completely submerged, it will not remain in that position but will return to the position in which AN is vertical.

Let a plane be drawn through AN, as placed, perpendicular to the surface of the fluid and cutting the segment of the paraboloid in the parabola BAB', the base of the segment in BB', and the plane of the surface of the fluid in the chord QQ' of the parabola.

Draw the tangent PT parallel to QQ', and the diameter PV, bisecting QQ', will accordingly be the axis of the portion of the paraboloid above the surface of the fluid.

Let F be the centre of gravity of the portion above the surface, C that of the whole solid, and produce FC to H, the centre of gravity of the immersed portion.

As in the last proposition, $AC > \frac{p}{2}$, and we measure CO along CA equal to $\frac{p}{2}$, and OR along OC equal to $\frac{1}{2}AO$.

Then $AN = \frac{3}{2}AC$, and $AR = \frac{3}{2}AO$;
and we derive, as before,
$$AR = (AN - \tfrac{3}{4}p).$$
Now, by hypothesis,
(spec. gravity of solid) : (spec. gravity of fluid)
$$\not> \{AN^2 - (AN - \tfrac{3}{4}p)^2\} : AN^2$$
$$\not> (AN^2 - AR^2) : AN^2.$$

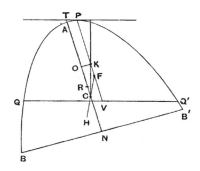

Therefore
(portion submerged) : (whole solid)
$$\not> (AN^2 - AR^2) : AN^2,$$
and   (whole solid) : (portion above surface)
$$\not> AN^2 : AR^2.$$
Thus $AN^2 : PV^2 \not> AN^2 : AR^2$,
whence $PV \not< AR$,
and $PF \not< \tfrac{2}{3}AR$
$$\not< AO.$$

Therefore, if a perpendicular to AC be drawn from O, it will meet PF in some point K between P and F.

And, since $CO = \tfrac{1}{2}p$, CK will be perpendicular to PT, as in the last proposition.

Now the force acting on the submerged portion of the solid will act upwards through H, and the weight of the other portion downwards through F, in directions parallel in both cases to CK; whence the proposition follows.

Proposition 6.

If a right segment of a paraboloid lighter than a fluid be such that its axis AM is greater than $\tfrac{3}{4}p$, but $AM : \tfrac{1}{2}p < 15 : 4$, and if the segment be placed in the fluid with its axis so inclined to the vertical that its base touches the fluid, it will never remain in such a position that the base touches the surface in one point only.

Suppose the segment of the paraboloid to be placed in the position described, and let the plane through the axis AM perpendicular to the surface of the fluid intersect the segment of the paraboloid in the parabolic segment BAB' and the plane of the surface of the fluid in BQ.

Take C on AM such that $AC = 2CM$ (or so that C is the centre of gravity of the segment of the paraboloid), and measure CK along CA such that
$$AM : CK = 15 : 4.$$

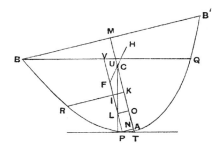

Thus $AM : CK > AM : \tfrac{1}{2}p$, by hypothesis; therefore $CK < \tfrac{1}{2}p$.

ON FLOATING BODIES II. 273

Measure CO along CA equal to $\frac{1}{2}p$. Also draw KR perpendicular to AC meeting the parabola in R.

Draw the tangent PT parallel to BQ, and through P draw the diameter PV bisecting BQ in V and meeting KR in I.

Then $\qquad PV : PI \underset{\text{or}>}{=} KM : AK,$

"*for this is proved.*"*

And $\qquad CK = \tfrac{4}{15}AM = \tfrac{2}{5}AC;$

whence $\qquad AK = AC - CK = \tfrac{3}{5}AC = \tfrac{2}{5}AM.$

Thus $\qquad KM = \tfrac{3}{5}AM.$

Therefore $\qquad KM = \tfrac{3}{2}AK.$

It follows that
$$PV \underset{\text{or}>}{=} \tfrac{3}{2}PI,$$

so that $\qquad PI \underset{\text{or}<}{=} 2IV.$

Let F be the centre of gravity of the immersed portion of the paraboloid, so that $PF = 2FV$. Produce FC to H, the centre of gravity of the portion above the surface.

Draw OL perpendicular to PV.

---

* We have no hint as to the work in which the proof of this proposition was contained. The following proof is shorter than Robertson's (in the Appendix to Torelli's edition).

Let BQ meet AM in U, and let PN be the ordinate from P to AM.

We have to prove that $PV . AK \underset{\text{or}>}{=} PI . KM$, or in other words that

$(PV . AK - PI . KM)$ is *positive* or *zero*.

Now $\qquad PV . AK - PI . KM = AK . PV - (AK - AN)(AM - AK)$
$\qquad\qquad\qquad\qquad\qquad = AK^2 - AK(AM + AN - PV) + AM . AN$
$\qquad\qquad\qquad\qquad\qquad = AK^2 - AK . UM + AM . AN,$

(since $AN = AT$).

Now $\qquad\qquad UM : BM = NT : PN.$

Therefore $\qquad UM^2 : p . AM = 4AN^2 : p . AN,$

whence $\qquad\qquad UM^2 = 4AM . AN,$

or $\qquad\qquad AM . AN = \dfrac{UM^2}{4}.$

Therefore $\qquad PV . AK - PI . KM = AK^2 - AK . UM + \dfrac{UM^2}{4}$
$\qquad\qquad\qquad\qquad\qquad = \left(AK - \dfrac{UM}{2}\right)^2,$

and accordingly $(PV . AK - PI . KM)$ cannot be negative.

Then, since $CO = \tfrac{1}{2}p$, CL must be perpendicular to PT and therefore to the surface of the fluid.

And the forces acting on the immersed portion of the paraboloid and the portion above the surface act respectively upwards and downwards along lines through F and H parallel to CL.

Hence the paraboloid cannot remain in the position in which B just touches the surface, but must turn in the direction of increasing the angle PTM.

The proof is the same in the case where the point I is not on VP but on VP produced, as in the second figure*.

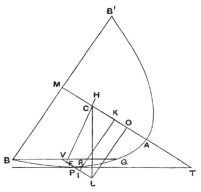

[With the notation used on p. 266, if the base BB' touch the surface of the fluid at B, we have
$$BM = BV \sin \theta + PN,$$
and, by the property of the parabola,
$$BV^2 = (p + 4AN) PV$$
$$= pk(1 + \cot^2 \theta).$$
Therefore $\quad \sqrt{ph} = \sqrt{pk} + \dfrac{p}{2} \cot \theta.$

To obtain the result of the proposition, we have to eliminate k between this equation and
$$x' - x = \cos \theta \left\{ \frac{p}{4}(\cot^2 \theta + 2) - \frac{2}{3}(h-k) \right\}.$$

* It is curious that the figures given by Torelli, Nizze and Heiberg are all incorrect, as they all make the point which I have called I lie on BQ instead of VP produced.

We have, from the first equation,
$$k = h - \sqrt{ph}\cot\theta + \frac{p}{4}\cot^2\theta,$$
or
$$h - k = \sqrt{ph}\cot\theta - \frac{p}{4}\cot^2\theta.$$
Therefore
$$x' - x = \cos\theta\left\{\frac{p}{4}(\cot^2\theta + 2) - \frac{2}{3}(\sqrt{ph}\cot\theta - \frac{p}{4}\cot^2\theta)\right\}$$
$$= \cos\theta\left\{\frac{p}{4}(\tfrac{5}{3}\cot^2\theta + 2) - \tfrac{2}{3}\sqrt{ph}\cot\theta\right\}.$$

If then the solid can never rest in the position described, but must turn in the direction of increasing the angle PTM, the expression within the bracket must be positive whatever be the value of θ.

Therefore $(\tfrac{2}{3})^2 ph < \tfrac{5}{6}p^2$,
or $h < \tfrac{15}{8}p$.]

Proposition 7.

Given a right segment of a paraboloid of revolution lighter than a fluid and such that its axis AM is greater than $\tfrac{3}{4}p$, but $AM : \tfrac{1}{2}p < 15 : 4$, if the segment be placed in the fluid so that its base is entirely submerged, it will never rest in such a position that the base touches the surface of the fluid at one point only.

Suppose the solid so placed that one point of the base only (B) touches the surface of the fluid. Let the plane through B and the axis AM cut the solid in the parabolic segment BAB' and the plane of the surface of the fluid in the chord BQ of the parabola.

Let C be the centre of gravity of the segment, so that $AC = 2CM$; and measure CK along CA such that
$$AM : CK = 15 : 4.$$
It follows that $CK < \tfrac{1}{2}p$.

Measure CO along CA equal to $\tfrac{1}{2}p$. Draw KR perpendicular to AM meeting the parabola in R.

Let PT, touching at P, be the tangent to the parabola which is parallel to BQ, and PV the diameter bisecting BQ, i.e. the axis of the portion of the paraboloid above the surface.

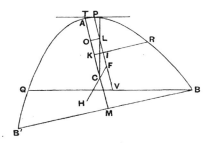

Then, as in the last proposition, we prove that

$$PV \underset{\text{or}>}{=} \tfrac{3}{2} PI,$$

and $$PI \underset{\text{or}<}{=} 2IV.$$

Let F be the centre of gravity of the portion of the solid above the surface; join FC and produce it to H, the centre of gravity of the portion submerged.

Draw OL perpendicular to PV; and, as before, since $CO = \tfrac{1}{2} p$, CL is perpendicular to the tangent PT. And the lines through H, F parallel to CL are perpendicular to the surface of the fluid; thus the proposition is established as before.

The proof is the same if the point I is not on VP but on VP produced.

Proposition 8.

Given a solid in the form of a right segment of a paraboloid of revolution whose axis AM is greater than $\tfrac{3}{4} p$, but such that $AM : \tfrac{1}{2} p < 15 : 4$, and whose specific gravity bears to that of a fluid a ratio less than $(AM - \tfrac{3}{4} p)^2 : AM^2$, then, if the solid be placed in the fluid so that its base does not touch the fluid and its axis is inclined at an angle to the vertical, the solid will not return to the position in which its axis is vertical and will not remain in any position except that in which its axis makes with the surface of the fluid a certain angle to be described.

Let am be taken equal to the axis AM, and let c be a point on am such that $ac = 2cm$. Measure co along ca equal to $\frac{1}{2}p$, and or along oc equal to $\frac{1}{2}ao$.

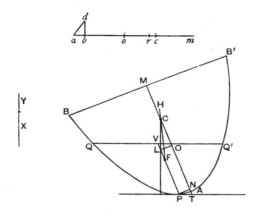

Let $X + Y$ be a straight line such that

(spec. gr. of solid) : (spec. gr. of fluid) $= (X + Y)^2 : am^2 \ldots\ldots(\alpha)$,

and suppose $X = 2Y$.

Now
$$ar = \tfrac{3}{2} ao = \tfrac{3}{2}(\tfrac{2}{3} am - \tfrac{1}{2}p)$$
$$= am - \tfrac{3}{4} p$$
$$= AM - \tfrac{3}{4} p.$$

Therefore, by hypothesis,
$$(X + Y)^2 : am^2 < ar^2 : am^2,$$
whence $(X + Y) < ar$, and therefore $X < ao$.

Measure ob along oa equal to X, and draw bd perpendicular to ab and of such length that
$$bd^2 = \tfrac{1}{2} co \cdot ab \ldots\ldots\ldots\ldots(\beta).$$
Join ad.

Now let the solid be placed in the fluid with its axis AM inclined at an angle to the vertical. Through AM draw a plane perpendicular to the surface of the fluid, and let this plane cut the paraboloid in the parabola BAB' and the plane of the surface of the fluid in the chord QQ' of the parabola.

Draw the tangent PT parallel to QQ', touching at P, and let PV be the diameter bisecting QQ' in V (or the axis of the immersed portion of the solid), and PN the ordinate from P.

Measure AO along AM equal to ao, and OC along OM equal to oc, and draw OL perpendicular to PV.

I. Suppose the angle OTP greater than the angle dab.

Thus $PN^2 : NT^2 > db^2 : ba^2$.

But $PN^2 : NT^2 = p : 4AN$
$= co : NT$,

and $db^2 : ba^2 = \frac{1}{2} co : ab$, by ($\beta$).

Therefore $NT < 2ab$,

or $AN < ab$,

whence $NO > bo$ (since $ao = AO$)
$> X$.

Now $(X+Y)^2 : am^2 =$ (spec. gr. of solid) : (spec. gr. of fluid)
$=$ (portion immersed) : (rest of solid)
$= PV^2 : AM^2$,

so that $X + Y = PV$.

But $PL (= NO) > X$
$> \frac{2}{3}(X + Y)$, since $X = 2Y$,
$> \frac{2}{3} PV$,

or $PV < \frac{3}{2} PL$,

and therefore $PL > 2LV$.

Take a point F on PV so that $PF = 2FV$, i.e. so that F is the centre of gravity of the immersed portion of the solid.

Also $AC = ac = \frac{2}{3} am = \frac{2}{3} AM$, and therefore C is the centre of gravity of the whole solid.

Join FC and produce it to H, the centre of gravity of the portion of the solid above the surface.

Now, since $CO = \frac{1}{2}p$, CL is perpendicular to the surface of the fluid; therefore so are the parallels to CL through F and H. But the force on the immersed portion acts upwards through F and that on the rest of the solid downwards through H.

Therefore the solid will not rest but turn in the direction of diminishing the angle MTP.

II. Suppose the angle OTP less than the angle dab. In this case, we shall have, instead of the above results, the following,
$$AN > ab,$$
$$NO < X.$$
Also $\qquad PV > \tfrac{3}{2}PL,$
and therefore $\qquad PL < 2LV.$

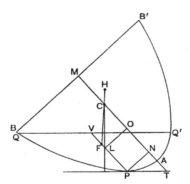

Make PF equal to $2FV$, so that F is the centre of gravity of the immersed portion.

And, proceeding as before, we prove in this case that the solid will turn in the direction of *increasing* the angle MTP.

III. When the angle MTP is equal to the angle dab, equalities replace inequalities in the results obtained, and L is itself the centre of gravity of the immersed portion. Thus all the forces act in one straight line, the perpendicular CL; therefore there is equilibrium, and the solid will rest in the position described.

[With the notation before used

$$x' - x = \cos\theta \left\{\frac{p}{4}(\cot^2\theta + 2) - \frac{2}{3}(h-k)\right\},$$

and a position of equilibrium is obtained by equating to zero the expression within the bracket. We have then

$$\frac{p}{4}\cot^2\theta = \frac{2}{3}(h-k) - \frac{p}{2}.$$

It is easy to verify that the angle θ satisfying this equation is the identical angle determined by Archimedes. For, in the above proposition,

$$\frac{3X}{2} = PV = k,$$

whence $\quad ab = \frac{2}{3}h - \frac{p}{2} - \frac{2}{3}k = \frac{2}{3}(h-k) - \frac{p}{2}.$

Also $\quad bd^2 = \frac{p}{4} \cdot ab.$

It follows that

$$\cot^2 dab = ab^2/bd^2 = \frac{4}{p}\left\{\frac{2}{3}(h-k) - \frac{p}{2}\right\}.]$$

Proposition 9.

Given a solid in the form of a right segment of a paraboloid of revolution whose axis AM is greater than $\frac{3}{4}p$, but such that $AM : \frac{1}{2}p < 15 : 4$, and whose specific gravity bears to that of a fluid a ratio greater than $\{AM^2 - (AM - \frac{3}{4}p)^2\} : AM^2$, then, if the solid be placed in the fluid with its axis inclined at an angle to the vertical but so that its base is entirely below the surface, the solid will not return to the position in which its axis is vertical and will not remain in any position except that in which its axis makes with the surface of the fluid an angle equal to that described in the last proposition.

Take am equal to AM, and take c on am such that $ac = 2cm$. Measure co along ca equal to $\frac{1}{2}p$, and ar along ac such that $ar = \frac{3}{2}ao$.

Let $X + Y$ be such a line that
(spec. gr. of solid) : (spec. gr. of fluid) $= \{am^2 - (X + Y)^2\} : am^2$,
and suppose $X = 2Y$.

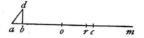

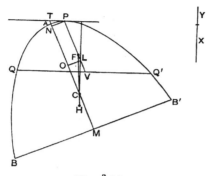

Now
$$ar = \tfrac{3}{2} ao$$
$$= \tfrac{3}{2} (\tfrac{2}{3} am - \tfrac{1}{2} p)$$
$$= AM - \tfrac{3}{4} p.$$

Therefore, by hypothesis,
$$am^2 - ar^2 : am^2 < \{am^2 - (X + Y)^2\} : am^2,$$
whence $\qquad X + Y < ar,$
and therefore $\qquad X < ao.$

Make ob (measured along oa) equal to X, and draw bd perpendicular to ba and of such length that
$$bd^2 = \tfrac{1}{2} co \cdot ab.$$

Join ad.

Now suppose the solid placed as in the figure with its axis AM inclined to the vertical. Let the plane through AM perpendicular to the surface of the fluid cut the solid in the parabola BAB' and the surface of the fluid in QQ'.

Let PT be the tangent parallel to QQ', PV the diameter bisecting QQ' (or the axis of the portion of the paraboloid above the surface), PN the ordinate from P.

I. Suppose the angle MTP greater than the angle dab. Let AM be cut as before in C and O so that $AC = 2CM$, $OC = \tfrac{1}{2}p$, and accordingly AM, am are equally divided. Draw OL perpendicular to PV.

Then, we have, as in the last proposition,
$$PN^2 : NT^2 > db^2 : ba^2,$$
whence $\qquad co : NT > \tfrac{1}{2}co : ab,$

and therefore $\qquad AN < ab.$

It follows that $\qquad NO > bo$
$$> X.$$

Again, since the specific gravity of the solid is to that of the fluid as the immersed portion of the solid to the whole,
$$AM^2 - (X+Y)^2 : AM^2 = AM^2 - PV^2 : AM^2,$$
or $\qquad (X+Y)^2 : AM^2 = PV^2 : AM^2.$

That is, $\qquad X + Y = PV.$

And $\qquad PL$ (or NO) $> X$
$$> \tfrac{2}{3} PV,$$
so that $\qquad PL > 2LV.$

Take F on PV so that $PF = 2FV$. Then F is the centre of gravity of the portion of the solid above the surface.

Also C is the centre of gravity of the whole solid. Join FC and produce it to H, the centre of gravity of the immersed portion.

Then, since $CO = \tfrac{1}{2}p$, CL is perpendicular to PT and to the surface of the fluid; and the force acting on the immersed portion of the solid acts upwards along the parallel to CL through H, while the weight of the rest of the solid acts downwards along the parallel to CL through F.

Hence the solid will not rest but turn in the direction of diminishing the angle MTP.

II. Exactly as in the last proposition, we prove that, if the angle MTP be less than the angle dab, the solid will not remain in its position but will turn in the direction of increasing the angle MTP.

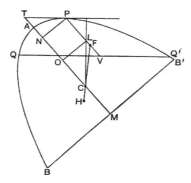

III. If the angle MTP is equal to the angle dab, the solid will rest in that position, because L and F will coincide, and all the forces will act along the one line CL.

Proposition 10.

Given a solid in the form of a right segment of a paraboloid of revolution in which the axis AM is of a length such that $AM : \tfrac{1}{2}p > 15 : 4$, and supposing the solid placed in a fluid of greater specific gravity so that its base is entirely above the surface of the fluid, to investigate the positions of rest.

(Preliminary.)

Suppose the segment of the paraboloid to be cut by a plane through its axis AM in the parabolic segment BAB_1 of which BB_1 is the base.

Divide AM at C so that $AC = 2CM$, and measure CK along CA so that
$$AM : CK = 15 : 4 \quad\ldots\ldots\ldots\ldots\ldots\ldots\ldots(\alpha),$$
whence, by the hypothesis, $CK > \tfrac{1}{2}p$.

Suppose CO measured along CA equal to $\tfrac{1}{2}p$, and take a point R on AM such that $MR = \tfrac{3}{2}CO$.

Thus
$$AR = AM - MR$$
$$= \tfrac{3}{2}(AC - CO)$$
$$= \tfrac{3}{2}AO.$$

Join BA, draw KA_2 perpendicular to AM meeting BA in A_2 bisect BA in A_3, and draw A_2M_2, A_3M_3 parallel to AM meeting BM in M_2, M_3 respectively.

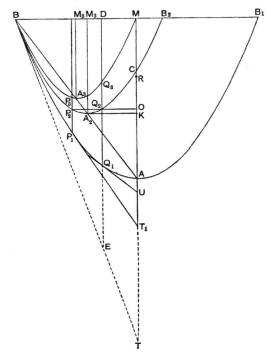

On A_2M_2, A_3M_3 as axes describe parabolic segments similar to the segment BAB_1. (It follows, by similar triangles, that BM will be the base of the segment whose axis is A_3M_3 and BB_2 the base of that whose axis is A_2M_2, where $BB_2 = 2BM_2$.)

The parabola BA_2B_2 will then pass through C.

[For $\quad BM_2 : M_2M = BM_2 : A_2K$

$\qquad = KM : AK$

$\qquad = CM + CK : AC - CK$

$\qquad = (\tfrac{1}{3} + \tfrac{4}{15})AM : (\tfrac{2}{3} - \tfrac{4}{15})AM$

$\qquad = 9 : 6 \dots\dots\dots\dots\dots\dots\dots\dots\dots\dots(\beta)$

$\qquad = MA : AC.$

Thus C is seen to be on the parabola BA_2B_2 by the converse of Prop. 4 of the *Quadrature of the Parabola*.]

Also, if a perpendicular to AM be drawn from O, it will meet the parabola BA_2B_2 in two points, as Q_2, P_2. Let $Q_1Q_2Q_3D$ be drawn through Q_2 parallel to AM meeting the parabolas BAB_1, BA_3M respectively in Q_1, Q_3 and BM in D; and let $P_1P_2P_3$ be the corresponding parallel to AM through P_2. Let the tangents to the outer parabola at P_1, Q_1 meet MA produced in T_1, U respectively.

Then, since the three parabolic segments are similar and similarly situated, with their bases in the same straight line and having one common extremity, and since $Q_1Q_2Q_3D$ is a diameter common to all three segments, it follows that

$$Q_1Q_2 : Q_2Q_3 = (B_2B_1 : B_1B).(BM : MB_2)^*.$$

Now $\qquad B_2B_1 : B_1B = MM_2 : BM \qquad$ (dividing by 2)

$\qquad\qquad\qquad = 2 : 5, \qquad$ by means of (β) above.

And $\qquad BM : MB_2 = BM : (2BM_2 - BM)$

$\qquad\qquad\qquad = 5 : (6 - 5), \qquad$ by means of (β),

$\qquad\qquad\qquad = 5 : 1.$

* This result is assumed without proof, no doubt as being an easy deduction from Prop. 5 of the *Quadrature of the Parabola*. It may be established as follows.

First, since AA_2A_3B is a straight line, and $AN=AT$ with the ordinary notation (where PT is the tangent at P and PN the ordinate), it follows, by similar triangles, that the tangent at B to the outer parabola is a tangent to each of the other two parabolas at the same point B.

Now, by the proposition quoted, if $DQ_3Q_2Q_1$ produced meet the tangent BT in E,

$$EQ_3 : Q_3D = BD : DM,$$

whence $\qquad EQ_3 : ED = BD : BM.$ ⎫

Similarly $\qquad EQ_2 : ED = BD : BB_2,$ ⎬

and $\qquad EQ_1 : ED = BD : BB_1.$ ⎭

The first two proportions are equivalent to

$$EQ_3 : ED = BD.BB_2 : BM.BB_2,$$

and $\qquad EQ_2 : ED = BD.BM : BM.BB_2.$

By subtraction,

$$Q_2Q_3 : ED = BD.MB_2 : BM.BB_2.$$

Similarly $\qquad Q_1Q_2 : ED = BD.B_2B_1 : BB_2.BB_1.$

It follows that

$$Q_1Q_2 : Q_2Q_3 = (B_2B_1 : B_1B).(BM : MB_2).$$

It follows that
$$Q_1Q_2 : Q_2Q_3 = 2 : 1,$$
or
$$Q_1Q_2 = 2Q_2Q_3.$$
Similarly
$$P_1P_2 = 2P_2P_3.$$
Also, since
$$MR = \tfrac{3}{2}CO = \tfrac{3}{4}p,$$
$$AR = AM - MR$$
$$= AM - \tfrac{3}{4}p.$$

(**Enunciation.**)

If the segment of the paraboloid be placed in the fluid with its base entirely above the surface, then

(I.) *if*

$$(spec.\ gr.\ of\ solid) : (spec.\ gr.\ of\ fluid) \not< AR^2 : AM^2$$
$$[\not< (AM - \tfrac{3}{4}p)^2 : AM^2],$$

the solid will rest in the position in which its axis AM is vertical;

(II.) *if*

$$(spec.\ gr.\ of\ solid) : (spec.\ gr.\ of\ fluid) < AR^2 : AM^2$$
$$but > Q_1Q_3^{\,2} : AM^2,$$

the solid will not rest with its base touching the surface of the fluid in one point only, but in such a position that its base does not touch the surface at any point and its axis makes with the surface an angle greater than U;

(III. a) *if*

$$(spec.\ gr.\ of\ solid) : (spec.\ gr.\ of\ fluid) = Q_1Q_3^{\,2} : AM^2,$$

the solid will rest and remain in the position in which the base touches the surface of the fluid at one point only and the axis makes with the surface an angle equal to U;

(III. b) *if*

$$(spec.\ gr.\ of\ solid) : (spec.\ gr.\ of\ fluid) = P_1P_3^{\,2} : AM^2,$$

the solid will rest with its base touching the surface of the fluid at one point only and with its axis inclined to the surface at an angle equal to T_1;

(IV.) *if*

$$(spec.\ gr.\ of\ solid) : (spec.\ gr.\ of\ fluid) > P_1P_3^{\,2} : AM^2$$
$$but < Q_1Q_3^{\,2} : AM^2,$$

the solid will rest and remain in a position with its base more submerged;

(V.) *if*

(spec. gr. of solid) : (spec. gr. of fluid) $< P_1P_3^2 : AM^2$,

the solid will rest in a position in which its axis is inclined to the surface of the fluid at an angle less than T_1, but so that the base does not even touch the surface at one point.

(**Proof.**)

(I.) Since $AM > \frac{3}{4}p$, and

(spec. gr. of solid) : (spec. gr. of fluid) $\not< (AM - \frac{3}{4}p)^2 : AM^2$,

it follows, by Prop. 4, that the solid will be in stable equilibrium with its axis vertical.

(II.) In this case

(spec. gr. of solid) : (spec. gr. of fluid) $< AR^2 : AM^2$

but $> Q_1Q_3^2 : AM^2$.

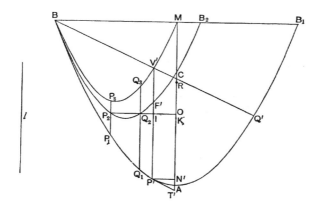

Suppose the ratio of the specific gravities to be equal to

$$l^2 : AM^2,$$

so that $l < AR$ but $> Q_1Q_3$.

Place $P'V'$ between the two parabolas BAB_1, BP_3Q_3M equal to l and parallel to AM*; and let $P'V'$ meet the intermediate parabola in F''.

Then, by the same proof as before, we obtain
$$P'F'' = 2F''V'.$$

Let $P'T'$, the tangent at P' to the outer parabola, meet MA in T', and let $P'N'$ be the ordinate at P'.

Join BV' and produce it to meet the outer parabola in Q'. Let OQ_2P_2 meet $P'V'$ in I.

Now, since, in two similar and similarly situated parabolic

* Archimedes does not give the solution of this problem, but it can be supplied as follows.

Let BR_1Q_1, BRQ_2 be two similar and similarly situated parabolic segments with their bases in the same straight line, and let BE be the common tangent at B.

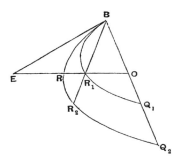

Suppose the problem solved, and let ERR_1O, parallel to the axes, meet the parabolas in R, R_1 and BQ_2 in O, making the intercept RR_1 equal to l.

Then, we have, as usual,
$$ER_1 : EO = BO : BQ_1$$
$$= BO \cdot BQ_2 : BQ_1 \cdot BQ_2,$$
and
$$ER : EO = BO : BQ_2$$
$$= BO \cdot BQ_1 : BQ_1 \cdot BQ_2.$$
By subtraction,
$$RR_1 : EO = BO \cdot Q_1Q_2 : BQ_1 \cdot BQ_2,$$
or
$$BO \cdot OE = l \cdot \frac{BQ_1 \cdot BQ_2}{Q_1Q_2}, \text{ which is known.}$$

And the ratio $BO : OE$ is known. Therefore BO^2, or OE^2, can be found, and therefore O.

ON FLOATING BODIES II. 289

segments with bases BM, BB_1 in the same straight line, BV', BQ' are drawn making the same angle with the bases,
$$BV' : BQ' = BM : BB_1 *$$
$$= 1 : 2,$$
so that $\qquad BV' = V'Q'.$

Suppose the segment of the paraboloid placed in the fluid, as described, with its axis inclined at an angle to the vertical, and with its base touching the surface at one point B only. Let the solid be cut by a plane through the axis and per-

pendicular to the surface of the fluid, and let the plane intersect the solid in the parabolic segment BAB' and the plane of the surface of the fluid in BQ.

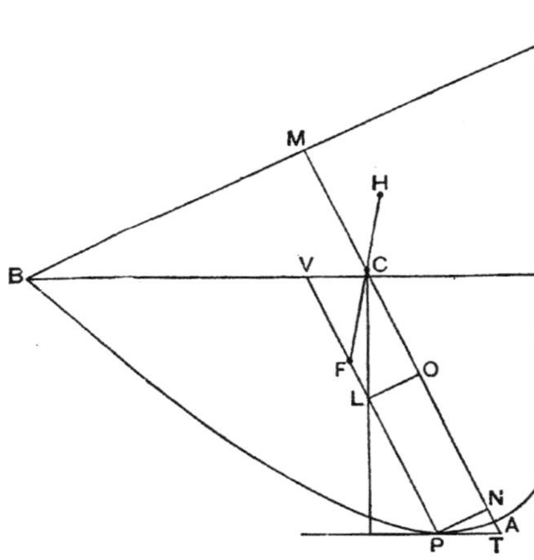

Take the points C, O on AM as before described. Draw

* To prove this, suppose that, in the figure on the opposite page, BR_1 is produced to meet the outer parabola in R_2.

We have, as before,
$$ER_1 : EO = BO : BQ_1,$$
$$ER : EO = BO : BQ_2,$$
whence $\qquad ER_1 : ER = BQ_2 : BQ_1.$

And, since R_1 is a point within the outer parabola,
$$ER : ER_1 = BR_1 : BR_2, \text{ in like manner.}$$
Hence $\qquad BQ_1 : BQ_2 = BR_1 : BR_2.$

the tangent parallel to BQ touching the parabola in P and meeting AM in T; and let PV be the diameter bisecting BQ (i.e. the axis of the immersed portion of the solid).

Then
$$l^2 : AM^2 = \text{(spec. gr. of solid)} : \text{(spec. gr. of fluid)}$$
$$= \text{(portion immersed)} : \text{(whole solid)}$$
$$= PV^2 : AM^2,$$
whence $\qquad P'V' = l = PV.$

Thus the segments in the two figures, namely $BP'Q'$, BPQ, are equal and similar.

Therefore $\qquad \angle PTN = \angle P'T'N'.$

Also $\qquad AT = AT',\ AN = AN',\ PN = P'N'.$

Now, in the first figure, $P'I < 2IV'.$

Therefore, if OL be perpendicular to PV in the second figure,
$$PL < 2LV.$$

Take F on LV so that $PF = 2FV$, i.e. so that F is the centre of gravity of the immersed portion of the solid. And C is the centre of gravity of the whole solid. Join FC and produce it to H, the centre of gravity of the portion above the surface.

Now, since $CO = \frac{1}{2}p$, CL is perpendicular to the tangent at P and to the surface of the fluid. Thus, as before, we prove that the solid will not rest with B touching the surface, but will turn in the direction of increasing the angle PTN.

Hence, in the position of rest, the axis AM must make with the surface of the fluid an angle greater than the angle U which the tangent at Q_1 makes with AM.

(III. a) In this case
$$\text{(spec. gr. of solid)} : \text{(spec. gr. of fluid)} = Q_1 Q_3^2 : AM^2.$$

Let the segment of the paraboloid be placed in the fluid so that its base nowhere touches the surface of the fluid, and its axis is inclined at an angle to the vertical.

ON FLOATING BODIES II. 291

Let the plane through AM perpendicular to the surface of the fluid cut the paraboloid in the parabola BAB' and the

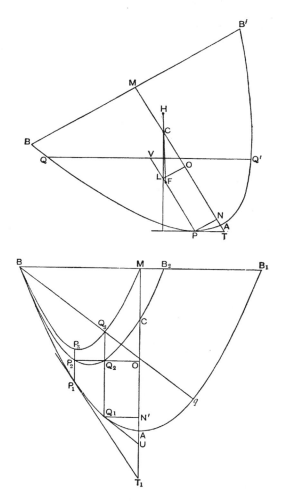

plane of the surface of the fluid in QQ'. Let PT be the tangent parallel to QQ', PV the diameter bisecting QQ', PN the ordinate at P.

Divide AM as before at C, O.

In the other figure let $Q_1 N'$ be the ordinate at Q_1. Join BQ_3 and produce it to meet the outer parabola in q. Then $BQ_3 = Q_3 q$, and the tangent $Q_1 U$ is parallel to Bq. Now

$$Q_1 Q_3{}^2 : AM^2 = \text{(spec. gr. of solid)} : \text{(spec. gr. of fluid)}$$
$$= \text{(portion immersed)} : \text{(whole solid)}$$
$$= PV^2 : AM^2.$$

Therefore $Q_1 Q_3 = PV$; and the segments QPQ', $BQ_1 q$ of the paraboloid are equal in volume. And the base of one passes through B, while the base of the other passes through Q, a point nearer to A than B is.

It follows that the angle between QQ' and BB' is less than the angle $B_1 Bq$.

Therefore $\quad\quad\quad \angle U < \angle PTN,$
whence $\quad\quad\quad\quad AN' > AN,$
and therefore $\quad\quad N'O$ (or $Q_1 Q_2$) $< PL,$
where OL is perpendicular to PV.

It follows, since $Q_1 Q_2 = 2 Q_2 Q_3$, that
$$PL > 2LV.$$

Therefore F, the centre of gravity of the immersed portion of the solid, is between P and L, while, as before, CL is perpendicular to the surface of the fluid.

Producing FC to H, the centre of gravity of the portion of the solid above the surface, we see that the solid must turn in the direction of diminishing the angle PTN until one point B of the base just touches the surface of the fluid.

When this is the case, we shall have a segment BPQ equal and similar to the segment $BQ_1 q$, the angle PTN will be equal to the angle U, and AN will be equal to AN'.

Hence in this case $PL = 2LV$, and F, L coincide, so that F, C, H are all in one vertical straight line.

Thus the paraboloid will remain in the position in which one point B of the base touches the surface of the fluid, and the axis makes with the surface an angle equal to U.

ON FLOATING BODIES II. 293

(III. b) In the case where

(spec. gr. of solid) : (spec. gr. of fluid) $= P_1P_3^2 : AM^2$,

we can prove in the same way that, if the solid be placed in the fluid so that its axis is inclined to the vertical and its base does not anywhere touch the surface of the fluid, the solid will take up and rest in the position in which one point only of the base touches the surface, and the axis is inclined to it at an angle equal to T_1 (in the figure on p. 284).

(IV.) In this case

(spec. gr. of solid) : (spec. gr. of fluid) $> P_1P_3^2 : AM^2$

but $< Q_1Q_3^2 : AM^2$.

Suppose the ratio to be equal to $l^2 : AM^2$, so that l is greater than P_1P_3 but less than Q_1Q_3.

Place $P'V'$ between the parabolas BP_1Q_1, BP_3Q_3 so that $P'V'$ is equal to l and parallel to AM, and let $P'V'$ meet the intermediate parabola in F' and OQ_2P_2 in I.

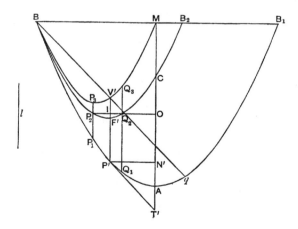

Join BV' and produce it to meet the outer parabola in q.

Then, as before, $BV' = V'q$, and accordingly the tangent $P'T'$ at P' is parallel to Bq. Let $P'N'$ be the ordinate of P'.

1. Now let the segment be placed in the fluid, *first*, with its axis so inclined to the vertical that its base does not anywhere touch the surface of the fluid.

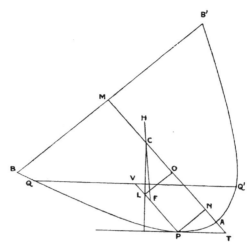

Let the plane through AM perpendicular to the surface of the fluid cut the paraboloid in the parabola BAB' and the plane of the surface of the fluid in QQ'. Let PT be the tangent parallel to QQ', PV the diameter bisecting QQ'. Divide AM at C, O as before, and draw OL perpendicular to PV.

Then, as before, we have $PV = l = P'V'$.

Thus the segments $BP'q$, QPQ' of the paraboloid are equal in volume; and it follows that the angle between QQ' and BB' is less than the angle B_1Bq.

Therefore $\angle P'T'N' < \angle PTN$,
and hence $AN' > AN$,
so that $NO > N'O$,
i.e. $PL > P'I$
$> P'F'$, a fortiori.

Thus $PL > 2LV$, so that F, the centre of gravity of the immersed portion of the solid, is between L and P, while CL is perpendicular to the surface of the fluid.

If then we produce FC to H, the centre of gravity of the portion of the solid above the surface, we prove that the solid will not rest but turn in the direction of diminishing the angle PTN.

2. Next let the paraboloid be so placed in the fluid that its base touches the surface of the fluid at one point B only, and let the construction proceed as before.

Then $PV = P'V'$, and the segments BPQ, $BP'q$ are equal and similar, so that
$$\angle PTN = \angle P'T'N'.$$
It follows that $\quad AN = AN',\ NO = N'O,$
and therefore $\quad P'I = PL,$
whence $\quad PL > 2LV.$

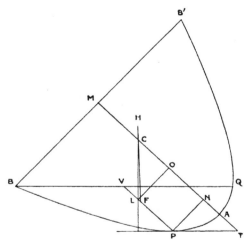

Thus F again lies between P and L, and, as before, the paraboloid will turn in the direction of diminishing the angle PTN, i.e. so that the base will be more submerged.

(V.) In this case

(spec. gr. of solid) : (spec. gr. of fluid) $< P_1P_3{}^2 : AM^2$.

If then the ratio is equal to $l^2 : AM^2$, $l < P_1P_3$. Place $P'V'$ between the parabolas BP_1Q_1 and BP_3Q_3 equal in length to l

and parallel to AM. Let $P'V'$ meet the intermediate parabola in F' and OP_2 in I.

Join BV' and produce it to meet the outer parabola in q. Then, as before, $BV' = V'q$, and the tangent $P'T'$ is parallel to Bq.

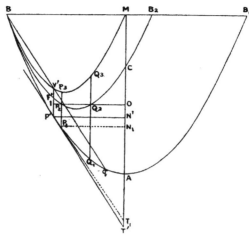

1. Let the paraboloid be so placed in the fluid that its base touches the surface at one point only.

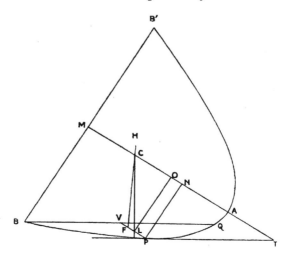

Let the plane through AM perpendicular to the surface of the fluid cut the paraboloid in the parabolic section BAB' and the plane of the surface of the fluid in BQ.

Making the usual construction, we find
$$PV = l = P'V',$$
and the segments BPQ, BP_1q are equal and similar.

Therefore $\angle PTN = \angle P'T'N'$,

and $AN = AN'$, $N'O = NO$.

Therefore $PL = P'I$,

whence it follows that $PL < 2LV$.

Thus F, the centre of gravity of the immersed portion of the solid, lies between L and V, while CL is perpendicular to the surface of the fluid.

Producing FC to H, the centre of gravity of the portion above the surface, we prove, as usual, that there will not be rest, but the solid will turn in the direction of increasing the angle PTN, so that the base will not anywhere touch the surface.

2. The solid will however rest in a position where its axis makes with the surface of the fluid an angle less than T_1.

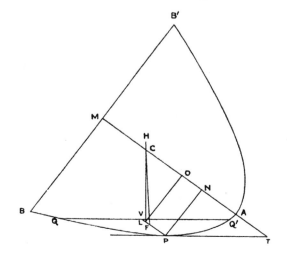

For let it be placed so that the angle PTN is not less than T_1.

Then, with the same construction as before, $PV = l = P'V'$.

And, since $\quad \angle T \not< \angle T_1,$

$\qquad\qquad AN \not> AN_1,$

and therefore $NO \not< N_1O$, where P_1N_1 is the ordinate of P_1.

Hence $\qquad\qquad PL \not< P_1P_2.$

But $\qquad\qquad P_1P_2 > P'F'.$

Therefore $\qquad\qquad PL > \tfrac{2}{3}PV,$

so that F, the centre of gravity of the immersed portion of the solid, lies between P and L.

Thus the solid will turn in the direction of diminishing the angle PTN until that angle becomes less than T_1.

[As before, if x, x' be the distances from T of the orthogonal projections of C, F respectively on TP, we have

$$x' - x = \cos\theta \left\{ \frac{p}{4}(\cot^2\theta + 2) - \frac{2}{3}(h - k) \right\} \ldots\ldots(1),$$

where $h = AM$, $k = PV$.

Also, if the base BB' touch the surface of the fluid at one point B, we have further, as in the note following Prop. 6,

$$\sqrt{ph} = \sqrt{pk} + \frac{p}{2}\cot\theta \ldots\ldots\ldots\ldots\ldots(2),$$

and $\qquad\qquad h - k = \sqrt{ph}\cot\theta - \frac{p}{4}\cot^2\theta \ \ldots\ldots\ldots\ (3).$

Therefore, to find the relation between h and the angle θ at which the axis of the paraboloid is inclined to the surface of the fluid in a position of equilibrium with B just touching the surface, we eliminate k and equate the expression in (1) to zero; thus

$$\frac{p}{4}(\cot^2\theta + 2) - \frac{2}{3}\left(\sqrt{ph}\cot\theta - \frac{p}{4}\cot^2\theta\right) = 0,$$

or $\qquad\qquad 5p\cot^2\theta - 8\sqrt{ph}\cot\theta + 6p = 0 \ \ldots\ldots\ldots\ (4).$

The two values of θ are given by the equations

$$5\sqrt{p}\cot\theta = 4\sqrt{h} \pm \sqrt{16h - 30p} \quad \ldots\ldots\ldots\ldots (5).$$

The lower sign corresponds to the angle U, and the upper sign to the angle T_1, in the proposition of Archimedes, as can be verified thus.

In the first figure of Archimedes (p. 284 above) we have

$$AK = \tfrac{2}{5}h,$$

$$M_2D^2 = \tfrac{3}{5}p \cdot OK = \tfrac{3}{5}p\,(\tfrac{2}{3}h - \tfrac{2}{5}h - \tfrac{1}{2}p)$$

$$= \frac{3p}{5}\left(\frac{4h}{15} - \frac{p}{2}\right).$$

If $P_1P_2P_3$ meet BM in D', it follows that

$$\left.\begin{array}{r}M_3D\\M_3D'\end{array}\right\} = M_2D \pm M_3M_2$$

$$= \sqrt{\frac{3p}{5}\left(\frac{4h}{15} - \frac{p}{2}\right)} \pm \frac{1}{10}\sqrt{ph},$$

and
$$\left.\begin{array}{r}MD\\MD'\end{array}\right\} = MM_2 \mp M_2D$$

$$= \frac{2}{5}\sqrt{ph} \mp \sqrt{\frac{3p}{5}\left(\frac{4h}{15} - \frac{p}{2}\right)}.$$

Now, from the property of the parabola,

$$\cot U = 2MD/p,$$
$$\cot T_1 = 2MD'/p,$$

so that
$$\frac{p}{2}\cot\left\{\begin{array}{c}U\\T_1\end{array}\right\} = \frac{2}{5}\sqrt{ph} \mp \sqrt{\frac{3p}{5}\left(\frac{4h}{15} - \frac{p}{2}\right)},$$

or
$$5\sqrt{p}\cot\left\{\begin{array}{c}U\\T_1\end{array}\right\} = 4\sqrt{h} \mp \sqrt{16h - 30p},$$

which agrees with the result (5) above.

To find the corresponding ratio of the specific gravities, or k^2/h^2, we have to use equations (2) and (5) and to express k in terms of h and p.

Equation (2) gives, on the substitution in it of the value of $\cot\theta$ contained in (5),

$$\sqrt{k} = \sqrt{h} - \tfrac{1}{10}(4\sqrt{h} \pm \sqrt{16h - 30p})$$
$$= \tfrac{3}{5}\sqrt{h} \mp \tfrac{1}{10}\sqrt{16h - 30p},$$

whence we obtain, by squaring,

$$k = \tfrac{13}{25}h - \tfrac{3}{10}p \mp \tfrac{3}{25}\sqrt{h(16h - 30p)} \ldots\ldots\ldots (6).$$

The lower sign corresponds to the angle U and the upper to the angle T_1, and, in order to verify the results of Archimedes, we have simply to show that the two values of k are equal to Q_1Q_3, P_1P_3 respectively.

Now it is easily seen that

$$Q_1Q_3 = h/2 - MD^2/p + 2M_3D^2/p,$$
$$P_1P_3 = h/2 - MD'^2/p + 2M_3D'^2/p.$$

Therefore, using the values of MD, MD', M_3D, M_3D' above found, we have

$$\left.\begin{array}{r}Q_1Q_3 \\ P_1P_3\end{array}\right\} = \frac{h}{2} + \frac{3}{5}\left(\frac{4h}{15} - \frac{p}{2}\right) - \frac{7h}{50} \pm \frac{6}{5}\sqrt{\frac{3h}{5}\left(\frac{4h}{15} - \frac{p}{2}\right)}$$
$$= \tfrac{13}{25}h - \tfrac{3}{10}p \pm \tfrac{3}{25}\sqrt{h(16h - 30p)},$$

which are the values of k given in (6) above.]

BOOK OF LEMMAS.

Proposition 1.

If two circles touch at A, and if BD, EF be parallel diameters in them, ADF is a straight line.

[The proof in the text only applies to the particular case where the diameters are perpendicular to the radius to the point of contact, but it is easily adapted to the more general case by one small change only.]

Let O, C be the centres of the circles, and let OC be joined and produced to A. Draw DH parallel to AO meeting OF in H.

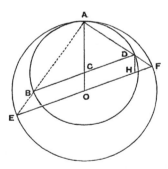

Then, since $OH = CD = CA$,
and $OF = OA$,
we have, by subtraction,
$HF = CO = DH$.
Therefore $\angle HDF = \angle HFD$.

Thus both the triangles CAD, HDF are isosceles, and the third angles ACD, DHF in each are equal. Therefore the equal angles in each are equal to one another, and

$$\angle ADC = \angle DFH.$$

Add to each the angle CDF, and it follows that

$$\angle ADC + \angle CDF = \angle CDF + \angle DFH$$
$$= \text{(two right angles)}.$$

Hence ADF is a straight line.

The same proof applies if the circles touch externally*.

Proposition 2.

Let AB be the diameter of a semicircle, and let the tangents to it at B and at any other point D on it meet in T. If now DE be drawn perpendicular to AB, and if AT, DE meet in F,

$$DF = FE.$$

Produce AD to meet BT produced in H. Then the angle ADB in the semicircle is right; therefore the angle BDH is also right. And TB, TD are equal.

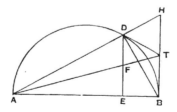

Therefore T is the centre of the semicircle on BH as diameter, which passes through D.

Hence $$HT = TB.$$

And, since DE, HB are parallel, it follows that $DF = FE$.

---

* Pappus assumes the result of this proposition in connexion with the ἄρβηλος (p. 214, ed. Hultsch), and he proves it for the case where the circles touch externally (p. 840).

BOOK OF LEMMAS. 303

Proposition 3.

Let P be any point on a segment of a circle whose base is AB, and let PN be perpendicular to AB. Take D on AB so that $AN = ND$. If now PQ be an arc equal to the arc PA, and BQ be joined,

BQ, BD shall be equal*.

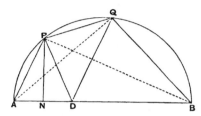

Join PA, PQ, PD, DQ.

* The segment in the figure of the ms. appears to have been a semicircle, though the proposition is equally true of any segment. But the case where the segment is a semicircle brings the proposition into close connexion with a proposition in Ptolemy's μεγάλη σύνταξις, I. 9 (p. 31, ed. Halma; cf. the reproduction in Cantor's *Gesch. d. Mathematik*, I. (1894), p. 389). Ptolemy's object is to connect by an equation the lengths of the chord of an arc and the chord of half the arc. Substantially his procedure is as follows. Suppose AP, PQ to be equal arcs, AB the diameter through A; and let AP, PQ, AQ, PB, QB be joined. Measure BD along BA equal to BQ. The perpendicular PN is now drawn, and it is proved that $PA = PD$, and $AN = ND$.

Then $AN = \tfrac{1}{2}(BA - BD) = \tfrac{1}{2}(BA - BQ) = \tfrac{1}{2}(BA - \sqrt{BA^2 - AQ^2})$.

And, by similar triangles, $AN : AP = AP : AB$.

Therefore $AP^2 = AB \cdot AN$

$= \tfrac{1}{2}(AB - \sqrt{AB^2 - AQ^2}) \cdot AB$.

This gives AP in terms of AQ and the known diameter AB. If we divide by AB^2 throughout, it is seen at once that the proposition gives a geometrical proof of the formula

$$\sin^2 \frac{a}{2} = \tfrac{1}{2}(1 - \cos a).$$

The case where the segment is a semicircle recalls also the method used by Archimedes at the beginning of the second part of Prop. 3 of the *Measurement of a circle*. It is there proved that, in the figure above,

$$AB + BQ : AQ = BP : PA,$$

or, if we divide the first two terms of the proposition by AB,

$$(1 + \cos a)/\sin a = \cot \frac{a}{2}.$$

Then, since the arcs PA, PQ are equal,
$$PA = PQ.$$
But, since $AN = ND$, and the angles at N are right,
$$PA = PD.$$
Therefore $PQ = PD$,
and $\angle PQD = \angle PDQ.$

Now, since A, P, Q, B are concyclic,
$$\angle PAD + \angle PQB = \text{(two right angles)},$$
whence $\angle PDA + \angle PQB = \text{(two right angles)}$
$$= \angle PDA + \angle PDB.$$
Therefore $\angle PQB = \angle PDB$;
and, since the parts, the angles PQD, PDQ, are equal,
$$\angle BQD = \angle BDQ,$$
and $BQ = BD.$

Proposition 4.

If AB be the diameter of a semicircle and N any point on AB, and if semicircles be described within the first semicircle and having AN, BN as diameters respectively, the figure included between the circumferences of the three semicircles is "what Archimedes called an ἄρβηλος"; and its area is equal to the circle on PN as diameter, where PN is perpendicular to AB and meets the original semicircle in P.*

For $AB^2 = AN^2 + NB^2 + 2AN \cdot NB$
$$= AN^2 + NB^2 + 2PN^2.$$

But circles (or semicircles) are to one another as the squares of their radii (or diameters).

---

* ἄρβηλος is literally 'a shoemaker's knife.' Cf. note attached to the remarks on the *Liber Assumptorum* in the Introduction, Chapter II.

Hence

(semicircle on AB) = (sum of semicircles on AN, NB)
  $+ 2$ (semicircle on PN).

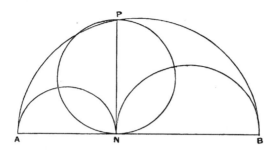

That is, the circle on PN as diameter is equal to the difference between the semicircle on AB and the sum of the semicircles on AN, NB, i.e. is equal to the area of the ἄρβηλος.

Proposition 5.

Let AB be the diameter of a semicircle, C any point on AB, and CD perpendicular to it, and let semicircles be described within the first semicircle and having AC, CB as diameters. Then, if two circles be drawn touching CD on different sides and each touching two of the semicircles, the circles so drawn will be equal.

Let one of the circles touch CD at E, the semicircle on AB in F, and the semicircle on AC in G.

Draw the diameter EH of the circle, which will accordingly be perpendicular to CD and therefore parallel to AB.

Join FH, HA, and FE, EB. Then, by Prop. 1, FHA, FEB are both straight lines, since EH, AB are parallel.

For the same reason AGE, CGH are straight lines.

Let AF produced meet CD in D, and let AE produced meet the outer semicircle in I. Join BI, ID.

Then, since the angles AFB, ACD are right, the straight lines AD, AB are such that the perpendiculars on each from the extremity of the other meet in the point E. Therefore, by the properties of triangles, AE is perpendicular to the line joining B to D.

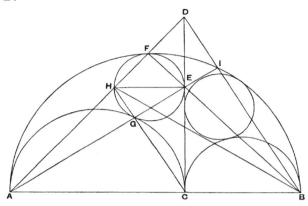

But AE is perpendicular to BI.

Therefore BID is a straight line.

Now, since the angles at G, I are right, CH is parallel to BD.

Therefore $\qquad AB : BC = AD : DH$
$\qquad\qquad\qquad\qquad = AC : HE,$
so that $\qquad AC . CB = AB . HE.$

In like manner, if d is the diameter of the other circle, we can prove that $\qquad AC . CB = AB . d.$

Therefore $d = HE$, and the circles are equal*.

---

* The property upon which this result depends, viz. that
$$AB : BC = AC : HE,$$
appears as an intermediate step in a proposition of Pappus (p. 230, ed. Hultsch) which proves that, in the figure above,
$$AB : BC = CE^2 : HE^2.$$

The truth of the latter proposition is easily seen. For, since the angle CEH is a right angle, and EG is perpendicular to CH,
$$CE^2 : EH^2 = CG : GH$$
$$= AC : HE.$$

[As pointed out by an Arabian Scholiast Alkauhi, this proposition may be stated more generally. If, instead of one point C on AB, we have two points C, D, and semicircles be described on AC, BD as diameters, and if, instead of the perpendicular to AB through C, we take the radical axis of the two semicircles, then the circles described on different sides of the radical axis and each touching it as well as two of the semicircles are equal. The proof is similar and presents no difficulty.]

Proposition 6.

Let AB, the diameter of a semicircle, be divided at C so that $AC = \frac{3}{2} CB$ [or in any ratio]. Describe semicircles within the first semicircle and on AC, CB as diameters, and suppose a circle drawn touching all three semicircles. If GH be the diameter of this circle, to find the relation between GH and AB.

Let GH be that diameter of the circle which is parallel to AB, and let the circle touch the semicircles on AB, AC, CB in D, E, F respectively.

Join AG, GD and BH, HD. Then, by Prop. 1, AGD, BHD are straight lines.

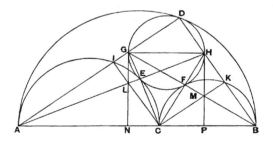

For a like reason AEH, BFG are straight lines, as also are CEG, CFH.

Let AD meet the semicircle on AC in I, and let BD meet the semicircle on CB in K. Join CI, CK meeting AE, BF

respectively in L, M, and let GL, HM produced meet AB in N, P respectively.

Now, in the triangle AGC, the perpendiculars from A, C on the opposite sides meet in L. Therefore, by the properties of triangles, GLN is perpendicular to AC.

Similarly HMP is perpendicular to CB.

Again, since the angles at I, K, D are right, CK is parallel to AD, and CI to BD.

Therefore $\quad AC : CB = AL : LH$
$\quad\quad\quad\quad\quad\quad\quad = AN : NP,$
and $\quad\quad BC : CA = BM : MG$
$\quad\quad\quad\quad\quad\quad\quad = BP : PN.$

Hence $\quad\quad AN : NP = NP : PB,$

or AN, NP, PB are in continued proportion*.

Now, in the case where $AC = \frac{3}{2} CB$,
$$AN = \frac{3}{2} NP = \frac{9}{4} PB,$$
whence $\quad BP : PN : NA : AB = 4 : 6 : 9 : 19.$

Therefore $\quad GH = NP = \frac{6}{19} AB.$

And similarly GH can be found when $AC : CB$ is equal to any other given ratio†.

* This same property appears incidentally in Pappus (p. 226) as an intermediate step in the proof of the "ancient proposition" alluded to below.

† In general, if $AC : CB = \lambda : 1$, we have
$$BP : PN : NA : AB = 1 : \lambda : \lambda^2 : (1 + \lambda + \lambda^2),$$
and $\quad\quad GH : AB = \lambda : (1 + \lambda + \lambda^2).$

It may be interesting to add the enunciation of the "ancient proposition" stated by Pappus (p. 208) and proved by him after several auxiliary lemmas.

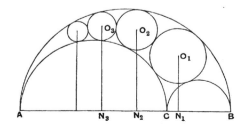

BOOK OF LEMMAS. 309

Proposition 7.

If circles be circumscribed about and inscribed in a square, the circumscribed circle is double of the inscribed circle.

For the ratio of the circumscribed to the inscribed circle is equal to that of the square on the diagonal to the square itself, i.e. to the ratio 2 : 1.

Proposition 8.

If AB be any chord of a circle whose centre is O, and if AB be produced to C so that BC is equal to the radius; if further CO meet the circle in D and be produced to meet the circle a second time in E, the arc AE will be equal to three times the arc BD.

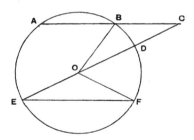

Draw the chord EF parallel to AB, and join OB, OF.

Let an ἄρβηλος be formed by three semicircles on AB, AC, CB as diameters, and let a series of circles be described, the first of which touches all three semicircles, while the second touches the first and two of the semicircles forming one end of the ἄρβηλος, the third touches the second and the same two semicircles, and so on. Let the diameters of the successive circles be d_1, d_2, d_3,... their centres O_1, O_2, O_3,... and O_1N_1, O_2N_2, O_3N_3,... the perpendiculars from the centres on AB. Then it is to be proved that

$$O_1N_1 = d_1,$$
$$O_2N_2 = 2d_2,$$
$$O_3N_3 = 3d_3,$$
$$\dots\dots\dots$$
$$O_nN_n = nd_n.$$

Then, since the angles OEF, OFE are equal,

$$\angle COF = 2 \angle OEF$$
$$= 2 \angle BCO, \text{ by parallels,}$$
$$= 2 \angle BOD, \text{ since } BC = BO.$$

Therefore
$$\angle BOF = 3 \angle BOD,$$

so that the arc BF is equal to three times the arc BD.

Hence the arc AE, which is equal to the arc BF, is equal to three times the arc BD*.

Proposition 9.

If in a circle two chords AB, CD which do not pass through the centre intersect at right angles, then

$$(arc\ AD) + (arc\ CB) = (arc\ AC) + (arc\ DB).$$

Let the chords intersect at O, and draw the diameter EF parallel to AB intersecting CD in H. EF will thus bisect CD at right angles in H, and

(arc ED) = (arc EC).

Also EDF, ECF are semicircles, while

(arc ED) = (arc EA) + (arc AD).

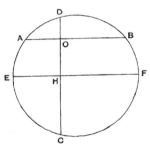

Therefore

(sum of arcs CF, EA, AD) = (arc of a semicircle).

And the arcs AE, BF are equal.

Therefore

(arc CB) + (arc AD) = (arc of a semicircle).

---

* This proposition gives a method of reducing the trisection of any angle, i.e. of any circular arc, to a problem of the kind known as νεύσεις. Suppose that AE is the arc to be trisected, and that ED is the diameter through E of the circle of which AE is an arc. In order then to find an arc equal to one-third of AE, we have only *to draw through A a line ABC, meeting the circle again in B and ED produced in C, such that BC is equal to the radius of the circle.* For a discussion of this and other νεύσεις see the Introduction, Chapter V.

Hence the remainder of the circumference, the sum of the arcs AC, DB, is also equal to a semicircle; and the proposition is proved.

Proposition 10.

Suppose that TA, TB are two tangents to a circle, while TC cuts it. Let BD be the chord through B parallel to TC, and let AD meet TC in E. Then, if EH be drawn perpendicular to BD, it will bisect it in H.

Let AB meet TC in F, and join BE.

Now the angle TAB is equal to the angle in the alternate segment, i.e.
$$\angle TAB = \angle ADB$$
$$= \angle AET, \text{ by parallels.}$$

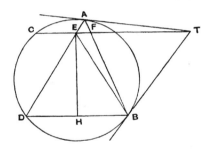

Hence the triangles EAT, AFT have one angle equal and another (at T) common. They are therefore similar, and
$$FT : AT = AT : ET.$$
Therefore
$$ET \cdot TF = TA^2$$
$$= TB^2.$$
It follows that the triangles EBT, BFT are similar.
Therefore
$$\angle TEB = \angle TBF$$
$$= \angle TAB.$$

But the angle TEB is equal to the angle EBD, and the angle TAB was proved equal to the angle EDB.

Therefore $\angle EDB = \angle EBD$.

And the angles at H are right angles.

It follows that $BH = HD$.*

Proposition 11.

If two chords AB, CD in a circle intersect at right angles in a point O, not being the centre, then

$$AO^2 + BO^2 + CO^2 + DO^2 = (diameter)^2.$$

Draw the diameter CE, and join AC, CB, AD, BE.

Then the angle CAO is equal to the angle CEB in the same segment, and the angles AOC, EBC are right; therefore the triangles AOC, EBC are similar, and

$$\angle ACO = \angle ECB.$$

It follows that the subtended arcs, and therefore the chords AD, BE, are equal.

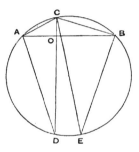

* The figure of this proposition curiously recalls the figure of a problem given by Pappus (pp. 836-8) among his lemmas to the first Book of the treatise of Apollonius *On Contacts* (περὶ ἐπαφῶν). The problem is, *Given a circle and two points E, F* (neither of which is necessarily, as in this case, the middle point of the chord of the circle drawn through E, F), *to draw through E, F respectively two chords AD, AB having a common extremity A and such that DB is parallel to EF.* The analysis is as follows. Suppose the problem solved, BD being parallel to FE. Let BT, the tangent at B, meet EF produced in T. (T is not in general the pole of AB, so that TA is not generally the tangent at A.)

Then $\angle TBF = \angle BDA$, in the alternate segment,

$= \angle AET$, by parallels.

Therefore A, E, B, T are concyclic, and

$$EF . FT = AF . FB.$$

But, the circle ADB and the point F being given, the rectangle $AF . FB$ is given. Also EF is given.

Hence FT is known.

Thus, to make the construction, we have only to find the length of FT from the data, produce EF to T so that FT has the ascertained length, draw the tangent TB, and then draw BD parallel to EF. DE, BF will then meet in A on the circle and will be the chords required.

Thus
$$(AO^2 + DO^2) + (BO^2 + CO^2) = AD^2 + BC^2$$
$$= BE^2 + BC^2$$
$$= CE^2.$$

Proposition 12.

If AB be the diameter of a semicircle, and TP, TQ the tangents to it from any point T, and if AQ, BP be joined meeting in R, then TR is perpendicular to AB.

Let TR produced meet AB in M, and join PA, QB.

Since the angle APB is right,
$$\angle PAB + \angle PBA = \text{(a right angle)}$$
$$= \angle AQB.$$

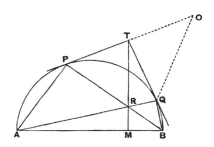

Add to each side the angle RBQ, and
$$\angle PAB + \angle QBA = \text{(exterior)} \angle PRQ.$$
But $\angle TPR = \angle PAB$, and $\angle TQR = \angle QBA$, in the alternate segments;
therefore $\angle TPR + \angle TQR = \angle PRQ.$

It follows from this that $TP = TQ = TR$.

[For, if PT be produced to O so that $TO = TQ$, we have
$$\angle TOQ = \angle TQO.$$
And, by hypothesis, $\angle PRQ = \angle TPR + TQR.$
By addition, $\angle POQ + \angle PRQ = \angle TPR + OQR.$

It follows that, in the quadrilateral $OPRQ$, the opposite angles are together equal to two right angles. Therefore a circle will go round $OPQR$, and T is its centre, because $TP = TO = TQ$. Therefore $TR = TP$.]

Thus $\angle TRP = \angle TPR = \angle PAM$.

Adding to each the angle PRM,

$$\angle PAM + \angle PRM = \angle TRP + \angle PRM$$
$$= \text{(two right angles)}.$$

Therefore $\angle APR + \angle AMR = \text{(two right angles)}$,

whence $\angle AMR = \text{(a right angle)}$*.

Proposition 13.

If a diameter AB of a circle meet any chord CD, not a diameter, in E, and if AM, BN be drawn perpendicular to CD, then
$$CN = DM\dagger.$$

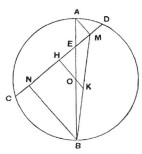

Let O be the centre of the circle, and OH perpendicular to CD. Join BM, and produce HO to meet BM in K.

Then $CH = HD$.

And, by parallels,

since $BO = OA$,

$BK = KM$.

Therefore $NH = HM$.

Accordingly $CN = DM$.

---

* TM is of course the polar of the intersection of PQ, AB, as it is the line joining the poles of PQ, AB respectively.

† This proposition is of course true whether M, N lie on CD or on CD produced each way. Pappus proves it for the latter case in his first lemma (p. 788) to the second Book of Apollonius' νεύσεις.

Proposition 14.

Let ACB be a semicircle on AB as diameter, and let AD, BE be equal lengths measured along AB from A, B respectively. On AD, BE as diameters describe semicircles on the side towards C, and on DE as diameter a semicircle on the opposite side. Let the perpendicular to AB through O, the centre of the first semicircle, meet the opposite semicircles in C, F respectively.

Then shall the area of the figure bounded by the circumferences of all the semicircles ("which Archimedes calls 'Salinon'"*) be equal to the area of the circle on CF as diameter†.

By Eucl. II. 10, since ED is bisected at O and produced to A,
$$EA^2 + AD^2 = 2(EO^2 + OA^2),$$
and $$CF = OA + OE = EA.$$

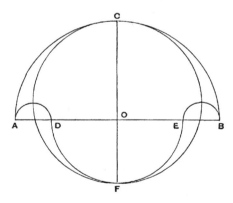

---

* For the explanation of this name see note attached to the remarks on the *Liber Assumptorum* in the Introduction, Chapter II. On the grounds there given at length I believe σάλινον to be simply a Graecised form of the Latin word *salinum*, 'salt-cellar.'

† Cantor (*Gesch. d. Mathematik*, I. p. 285) compares this proposition with Hippocrates' attempt to square the circle by means of lunes, but points out that the object of Archimedes may have been the converse of that of Hippocrates. For, whereas Hippocrates wished to find the area of a circle from that of other figures of the same sort, Archimedes' intention was possibly to equate the area of figures bounded by different curves to that of a circle regarded as already known.

316   ARCHIMEDES

Therefore
$$AB^2 + DE^2 = 4(EO^2 + OA^2) = 2(CF^2 + AD^2).$$

But circles (and therefore semicircles) are to one another as the squares on their radii (or diameters).

Therefore

(sum of semicircles on AB, DE)

    = (circle on CF) + (sum of semicircles on AD, BE).

Therefore

(area of 'salinon') = (area of circle on CF as diam.).

Proposition 15.

Let AB be the diameter of a circle, AC a side of an inscribed regular pentagon, D the middle point of the arc AC. Join CD and produce it to meet BA produced in E; join AC, DB meeting in F, and draw FM perpendicular to AB. Then

$$EM = (radius\ of\ circle)^*.$$

Let O be the centre of the circle, and join DA, DM, DO, CB.

Now $\angle ABC = \frac{2}{5}$ (right angle),

and $\angle ABD = \angle DBC = \frac{1}{5}$ (right angle),

whence $\angle AOD = \frac{2}{5}$ (right angle).

---

* Pappus gives (p. 418) a proposition almost identical with this among the lemmas required for the comparison of the five regular polyhedra. His enunciation is substantially as follows. If DH be half the side of a pentagon inscribed in a circle, while DH is perpendicular to the radius OHA, and if HM be made equal to AH, then OA is divided at M in extreme and mean ratio, OM being the greater segment.

In the course of the proof it is first shown that AD, DM, MO are all equal, as in the proposition above.

Then, the triangles ODA, DAM being similar,

$$OA : AD = AD : AM,$$

or (since $AD = OM$) $OA : OM = OM : MA$.

BOOK OF LEMMAS. 317

Further, the triangles *FCB*, *FMB* are equal in all respects.

Therefore, in the triangles *DCB*, *DMB*, the sides *CB*, *MB* being equal and *BD* common, while the angles *CBD*, *MBD* are equal,

$$\angle BCD = \angle BMD = \tfrac{6}{5}\,(\text{right angle}).$$

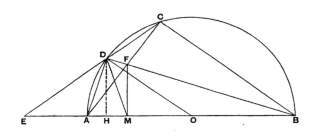

But $\angle BCD + \angle BAD = (\text{two right angles})$
$= \angle BAD + \angle DAE$
$= \angle BMD + \angle DMA,$

so that $\angle DAE = \angle BCD,$

and $\angle BAD = \angle AMD.$

Therefore $AD = MD.$

Now, in the triangle *DMO*,

$$\angle MOD = \tfrac{2}{5}\,(\text{right angle}),$$
$$\angle DMO = \tfrac{6}{5}\,(\text{right angle}).$$

Therefore $\angle ODM = \tfrac{2}{5}\,(\text{right angle}) = AOD;$

whence $OM = MD.$

Again $\angle EDA = (\text{supplement of } ADC)$
$= \angle CBA$
$= \tfrac{2}{5}\,(\text{right angle})$
$= \angle ODM.$

Therefore, in the triangles EDA, ODM,

$$\angle EDA = \angle ODM,$$
$$\angle EAD = \angle OMD,$$

and the sides AD, MD are equal.

Hence the triangles are equal in all respects, and

$$EA = MO.$$

Therefore $EM = AO.$

Moreover $DE = DO$; and it follows that, since DE is equal to the side of an inscribed hexagon, and DC is the side of an inscribed decagon, EC is divided at D in extreme and mean ratio [i.e. $EC : ED = ED : DC$]; "and this is proved in the book of the Elements." [Eucl. XIII. 9, "If the side of the hexagon and the side of the decagon inscribed in the same circle be put together, the whole straight line is divided in extreme and mean ratio, and the greater segment is the side of the hexagon."]

THE CATTLE-PROBLEM.

It is required to find the number of bulls and cows of each of four colours, or to find 8 unknown quantities. The first part of the problem connects the unknowns by seven simple equations; and the second part adds two more conditions to which the unknowns must be subject.

Let W, w be the numbers of white bulls and cows respectively,

$\quad\quad X, x$,, ,, black ,, ,, ,,
$\quad\quad Y, y$,, ,, yellow ,, ,, ,,
$\quad\quad Z, z$,, ,, dappled ,, ,, ,,

First part.

(I) $\quad\quad W = (\tfrac{1}{2} + \tfrac{1}{3}) X + Y \dots\dots\dots\dots\dots(\alpha)$,

$\quad\quad\quad\quad X = (\tfrac{1}{4} + \tfrac{1}{5}) Z + Y \dots\dots\dots\dots\dots(\beta)$,

$\quad\quad\quad\quad Z = (\tfrac{1}{6} + \tfrac{1}{7}) W + Y \dots\dots\dots\dots\dots(\gamma)$,

(II) $\quad\quad w = (\tfrac{1}{3} + \tfrac{1}{4})(X + x)\dots\dots\dots\dots\dots(\delta)$,

$\quad\quad\quad\quad x = (\tfrac{1}{4} + \tfrac{1}{5})(Z + z)\dots\dots\dots\dots\dots(\epsilon)$,

$\quad\quad\quad\quad z = (\tfrac{1}{5} + \tfrac{1}{6})(Y + y)\dots\dots\dots\dots\dots(\zeta)$,

$\quad\quad\quad\quad y = (\tfrac{1}{6} + \tfrac{1}{7})(W + w)\dots\dots\dots\dots\dots(\eta)$.

Second part.

$\quad\quad\quad\quad W + X = \text{a square}\dots\dots\dots\dots\dots\dots(\theta)$,

$\quad\quad\quad\quad Y + Z = \text{a triangular number}\dots\dots\dots\dots(\iota)$.

[There is an ambiguity in the language which expresses the condition (θ). Literally the lines mean "When the white bulls joined in number with the black, they stood firm (ἔμπεδον) with depth and breadth of equal measurement (ἰσόμετροι εἰς βάθος εἰς εὖρός τε); and the plains of Thrinakia, far-stretching all ways, were filled with their multitude" (reading, with Krumbiegel, πλήθους instead of πλίνθου). Considering that, if the bulls were packed together so as to form a square *figure*, the number of them need not be a square *number*, since a bull is longer than it is broad, it is clear that one possible interpretation would be to take the 'square' to be a square *figure*, and to understand condition (θ) to be simply

$W + X =$ a rectangle (i.e. a product of two factors).

The problem may therefore be stated in two forms:

(1) the simpler one in which, for the condition (θ), there is substituted the mere requirement that

$W + X =$ a product of two whole numbers;

(2) the complete problem in which all the conditions have to be satisfied including the requirement (θ) that

$W + X =$ a square number.

The simpler problem was solved by Jul. Fr. Wurm and may be called

Wurm's Problem.

The solution of this is given (together with a discussion of the complete problem) by Amthor in the *Zeitschrift für Math. u. Physik* (*Hist. litt. Abtheilung*), XXV. (1880), p. 156 sqq.

Multiply (α) by 336, (β) by 280, (γ) by 126, and add; thus

$297 W = 742 Y$, or $3^3 . 11 W = 2 . 7 . 53 Y$ ....(α').

Then from (γ) and (β) we obtain

$891 Z = 1580 Y$, or $3^4 . 11 Z = 2^2 . 5 . 79 Y$ ......(β'),

and $99 X = 178 Y$, or $3^2 . 11 X = 2 . 89 Y$ ....(γ').

Again, if we multiply (δ) by 4800, (ϵ) by 2800, (ζ) by 1260, (η) by 462, and add, we obtain

$4657 w = 2800 X + 1260 Z + 462 Y + 143 W$;

THE CATTLE-PROBLEM. 321

and, by means of the values in (α'), (β'), (γ'), we derive

$$297 \cdot 4657w = 2402120 Y,$$

or $\quad\quad 3^3 \cdot 11 \cdot 4657w = 2^3 \cdot 5 \cdot 7 \cdot 23 \cdot 373 Y \quad\ldots\ldots(\delta').$

Hence, by means of (η), (ζ), (ϵ), we have

$$3^2 \cdot 11 \cdot 4657y = 13 \cdot 46489 Y \ldots\ldots\ldots\ldots(\epsilon'),$$

$$3^3 \cdot 4657z = 2^2 \cdot 5 \cdot 7 \cdot 761 Y \ldots\ldots\ldots\ldots(\zeta'),$$

and $\quad\quad 3^2 \cdot 11 \cdot 4657x = 2 \cdot 17 \cdot 15991 Y \ldots\ldots\ldots\ldots(\eta').$

And, since all the unknowns must be whole numbers, we see from the equations (α'), (β'), ... (η') that Y must be divisible by $3^4 \cdot 11 \cdot 4657$, i.e. we may put

$$Y = 3^4 \cdot 11 \cdot 4657n = 4149387n.$$

Therefore the equations (α'), (β'),...(η') give the following values for all the unknowns in terms of n, viz.

$$\left.\begin{array}{l}W = 2 \cdot 3 \cdot 7 \cdot 53 \cdot 4657n \quad = 10366482n \\ X = 2 \cdot 3^2 \cdot 89 \cdot 4657n \quad = 7460514n \\ Y = 3^4 \cdot 11 \cdot 4657n \quad = 4149387n \\ Z = 2^2 \cdot 5 \cdot 79 \cdot 4657n \quad = 7358060n \\ w = 2^3 \cdot 3 \cdot 5 \cdot 7 \cdot 23 \cdot 373n = 7206360n \\ x = 2 \cdot 3^2 \cdot 17 \cdot 15991n \quad = 4893246n \\ y = 3^2 \cdot 13 \cdot 46489n \quad = 5439213n \\ z = 2^2 \cdot 3 \cdot 5 \cdot 7 \cdot 11 \cdot 761n = 3515820n\end{array}\right\}\ldots\ldots\ldots(\text{A}).$$

If now $n = 1$, the numbers are the smallest which will satisfy the seven equations (α), (β),...(η); and we have next to find such an integral value for n that the equation (ι) will be satisfied also. [The modified equation (θ) requiring that $W + X$ must be a product of two factors is then simultaneously satisfied.]

Equation (ι) requires that

$$Y + Z = \frac{q(q+1)}{2},$$

where q is some positive integer.

Putting for Y, Z their values as above ascertained, we have

$$\frac{q(q+1)}{2} = (3^4 \cdot 11 + 2^2 \cdot 5 \cdot 79) \cdot 4657n$$

$$= 2471 \cdot 4657n$$

$$= 7 \cdot 353 \cdot 4657n.$$

Now q is either even or odd, so that either $q = 2s$, or $q = 2s - 1$, and the equation becomes

$$s(2s \pm 1) = 7 \cdot 353 \cdot 4657n.$$

As n need not be a prime number, we suppose $n = u \cdot v$, where u is the factor in n which divides s without a remainder and v the factor which divides $2s \pm 1$ without a remainder; we then have the following sixteen alternative pairs of simultaneous equations:

(1) $s = u,$ $2s \pm 1 = 7 \cdot 353 \cdot 4657v,$

(2) $s = 7u,$ $2s \pm 1 = 353 \cdot 4657v,$

(3) $s = 353u,$ $2s \pm 1 = 7 \cdot 4657v,$

(4) $s = 4657u,$ $2s \pm 1 = 7 \cdot 353v,$

(5) $s = 7 \cdot 353u,$ $2s \pm 1 = 4657v,$

(6) $s = 7 \cdot 4657u,$ $2s \pm 1 = 353v,$

(7) $s = 353 \cdot 4657u,$ $2s \pm 1 = 7v,$

(8) $s = 7 \cdot 353 \cdot 4657u,$ $2s \pm 1 = v.$

In order to find the least value of n which satisfies all the conditions of the problem, we have to choose from the various positive integral solutions of these pairs of equations that particular one which gives the smallest value for the product uv or n.

If we solve the various pairs and compare the results, we find that it is the pair of equations

$$s = 7u, \quad 2s - 1 = 353 \cdot 4657v,$$

which leads to the solution we want; this solution is then

$$u = 117423, \quad v = 1,$$

so that $n = uv = 117423 = 3^3 \cdot 4349,$

THE CATTLE-PROBLEM. 323

whence it follows that
$$s = 7u = 821961,$$
and
$$q = 2s - 1 = 1643921.$$
Thus
$$Y + Z = 2471 \,.\, 4657n$$
$$= 2471 \,.\, 4657 \,.\, 117423$$
$$= 1351238949081$$
$$= \frac{1643921 \,.\, 1643922}{2},$$

which is a triangular number, as required.

The number in equation (θ) which has to be the product of two integers is now

$$W + X = 2 \,.\, 3 \,.\, (7 \,.\, 53 + 3 \,.\, 89) \,.\, 4657n$$
$$= 2^2 \,.\, 3 \,.\, 11 \,.\, 29 \,.\, 4657n$$
$$= 2^2 \,.\, 3 \,.\, 11 \,.\, 29 \,.\, 4657 \,.\, 117423$$
$$= 2^2 \,.\, 3^4 \,.\, 11 \,.\, 29 \,.\, 4657 \,.\, 4349$$
$$= (2^2 \,.\, 3^4 \,.\, 4349) \,.\, (11 \,.\, 29 \,.\, 4657)$$
$$= 1409076 \,.\, 1485583,$$

which is a rectangular number with nearly equal factors.

The solution is then as follows (substituting for n its value 117423):

$$W = 1217263415886$$
$$X = 876035935422$$
$$Y = 487233469701$$
$$Z = 864005479380$$
$$w = 846192410280$$
$$x = 574579625058$$
$$y = 638688708099$$
$$\underline{z = 412838131860}$$
$$\text{and the sum} = 5916837175686$$

The complete problem.

In this case the seven original equations $(\alpha), (\beta),\ldots(\eta)$ have to be satisfied, and the following further conditions must hold,

$W + X =$ a square number $= p^2$, say,

$Y + Z =$ a triangular number $= \dfrac{q(q+1)}{2}$, say.

Using the values found above (A), we have in the first place

$$p^2 = 2 \cdot 3 \cdot (7 \cdot 53 + 3 \cdot 89) \cdot 4657n$$
$$= 2^2 \cdot 3 \cdot 11 \cdot 29 \cdot 4657n,$$

and this equation will be satisfied if

$$n = 3 \cdot 11 \cdot 29 \cdot 4657 \xi^2 = 4456749 \xi^2,$$

where ξ is any integer.

Thus the first 8 equations $(\alpha), (\beta),\ldots(\eta), (\theta)$ are satisfied by the following values:

$W = 2 \cdot 3^2 \cdot 7 \cdot 11 \cdot 29 \cdot 53 \cdot 4657^2 \cdot \xi^2 \qquad = 46200808287018 \cdot \xi^2$

$X = 2 \cdot 3^3 \cdot 11 \cdot 29 \cdot 89 \cdot 4657^2 \cdot \xi^2 \qquad = 33249638308986 \cdot \xi^2$

$Y = 3^5 \cdot 11^2 \cdot 29 \cdot 4657^2 \cdot \xi^2 \qquad = 18492776362863 \cdot \xi^2$

$Z = 2^2 \cdot 3 \cdot 5 \cdot 11 \cdot 29 \cdot 79 \cdot 4657^2 \cdot \xi^2 \qquad = 32793026546940 \cdot \xi^2$

$w = 2^3 \cdot 3^2 \cdot 5 \cdot 7 \cdot 11 \cdot 23 \cdot 29 \cdot 373 \cdot 4657 \cdot \xi^2 = 32116937723640 \cdot \xi^2$

$x = 2 \cdot 3^3 \cdot 11 \cdot 17 \cdot 29 \cdot 15991 \cdot 4657 \cdot \xi^2 \quad = 21807969217254 \cdot \xi^2$

$y = 3^3 \cdot 11 \cdot 13 \cdot 29 \cdot 46489 \cdot 4657 \cdot \xi^2 \quad = 24241207098537 \cdot \xi^2$

$z = 2^2 \cdot 3^2 \cdot 5 \cdot 7 \cdot 11^2 \cdot 29 \cdot 761 \cdot 4657 \cdot \xi^2 \quad = 15669127269180 \cdot \xi^2$

It remains to determine ξ so that equation (ι) may be satisfied, i.e. so that

$$Y + Z = \dfrac{q(q+1)}{2}.$$

Substituting the ascertained values of Y, Z, we have

$$\dfrac{q(q+1)}{2} = 51285802909803 \cdot \xi^2$$

$$= 3 \cdot 7 \cdot 11 \cdot 29 \cdot 353 \cdot 4657^2 \cdot \xi^2.$$

Multiply by 8, and put

$$2q + 1 = t, \quad 2 \cdot 4657 \cdot \xi = u,$$

and we have the "Pellian" equation

$$t^2 - 1 = 2 \cdot 3 \cdot 7 \cdot 11 \cdot 29 \cdot 353 \cdot u^2,$$

that is, $\quad t^2 - 4729494\, u^2 = 1.$

Of the solutions of this equation the smallest has to be chosen for which u is divisible by $2 \cdot 4657$.

When this is done,

$$\xi = \frac{u}{2 \cdot 4657} \text{ and is a whole number;}$$

whence, by substitution of the value of ξ so found in the last system of equations, we should arrive at the solution of the complete problem.

It would require too much space to enter on the solution of the "Pellian" equation

$$t^2 - 4729494\, u^2 = 1,$$

and the curious reader is referred to Amthor's paper itself. Suffice it to say that he develops $\sqrt{4729494}$ in the form of a continued fraction as far as the period which occurs after 91 convergents, and, after an arduous piece of work, arrives at the conclusion that

$$W = 1598 \;\langle 206541 \rangle,$$

where $\langle 206541 \rangle$ represents the fact that there are 206541 more digits to follow, and that, with the same notation,

the whole number of cattle $= 7766 \;\langle 206541 \rangle.$

One may well be excused for doubting whether Archimedes solved the complete problem, having regard to the enormous size of the numbers and the great difficulties inherent in the work. By way of giving an idea of the space which would be required for merely writing down the results when obtained, Amthor remarks that the large seven-figured logarithmic tables contain on one page 50 lines with 50 figures or so in each, say altogether 2500 figures; therefore *one* of the eight unknown quantities would, when found, occupy $82\frac{1}{2}$ such pages, and to write down all the eight numbers would require a volume of 660 pages!]

THE
METHOD OF ARCHIMEDES
RECENTLY DISCOVERED BY HEIBERG

A SUPPLEMENT TO *THE WORKS*
OF ARCHIMEDES 1897

EDITED BY

Sir THOMAS L. HEATH,
K.C.B., Sc.D., F.R.S.
SOMETIME FELLOW OF TRINITY COLLEGE, CAMBRIDGE

Cambridge:
at the University Press
1912

INTRODUCTORY NOTE

FROM the point of view of the student of Greek mathematics there has been, in recent years, no event comparable in interest with the discovery by Heiberg in 1906 of a Greek MS. containing, among other works of Archimedes, substantially the whole of a treatise which was formerly thought to be irretrievably lost.

The full description of the MS. as given in the preface to Vol. I. (1910) of the new edition of Heiberg's text of Archimedes now in course of publication is—

Codex rescriptus Metochii Constantinopolitani S. Sepulchri monasterii Hierosolymitani 355, 4to.

Heiberg has told the story of his discovery of this MS. and given a full description of it*. His attention having been called to a notice in Vol. IV. (1899) of the Ἱεροσολυμιτικὴ βιβλιοθήκη of Papadopulos Kerameus relating to a palimpsest of mathematical content, he at once inferred from a few specimen lines which were quoted that the MS. must contain something by Archimedes. As the result of inspection, at Constantinople, of the MS. itself, and by means of a photograph taken of it, he was able to see what it contained and to decipher much of the contents. This was in the year 1906, and he inspected the MS. once more in 1908. With the exception of the last leaves, 178 to 185, which are of paper of the 16th century, the MS. is of parchment and contains writings of Archimedes copied in a good hand of the 10th centuιy, in two columns. An attempt was made (fortunately with only partial success) to wash out the old writing, and then the parchment was used again, for the purpose of writing a Euchologion thereon, in the 12th—13th or 13th—14th centuries. The earlier writing appears with more or less clearness on most of the 177 leaves; only 29 leaves are destitute of any trace of such writing; from 9 more it was hopelessly washed off; on a few more leaves only a few words can be made out; and again some 14 leaves have old writing

* *Hermes* XLII. 1907, pp. 235 sq.

upon them in a different hand and with no division into columns. All the rest is tolerably legible with the aid of a magnifying glass. Of the treatises of Archimedes which are found in other MSS., the new MS. contains, in great part, the books *On the Sphere and Cylinder*, almost the whole of the work *On Spirals*, and some parts of the *Measurement of a Circle* and of the books *On the Equilibrium of Planes*. But the important fact is that it contains (1) a considerable proportion of the work *On Floating Bodies* which was formerly supposed to be lost so far as the Greek text is concerned and only to have survived in the translation by Wilhelm von Mörbeke, and (2), most precious of all, the greater part of the book called, according to its own heading, Ἔφοδος and elsewhere, alternatively, Ἐφόδιον or Ἐφοδικόν, meaning *Method*. The portion of this latter work contained in the MS. has already been published by Heiberg (1) in Greek* and (2) in a German translation with commentary by Zeuthen †. The treatise was formerly only known by an allusion to it in Suidas, who says that Theodosius wrote a commentary upon it; but the *Metrica* of Heron, newly discovered by R. Schöne and published in 1903, quotes three propositions from it‡, including the two main propositions enunciated by Archimedes at the beginning as theorems novel in character which the method furnished a means of investigating. Lastly the MS. contains two short propositions, in addition to the preface, of a work called *Stomachion* (as it might be "Neck-Spiel" or "Quäl-Geist") which treated of a sort of Chinese puzzle known afterwards by the name of "loculus Archimedius"; it thus turns out that this puzzle, which Heiberg was formerly disinclined to attribute to Archimedes§, is really genuine.

The *Method*, so happily recovered, is of the greatest interest for the following reason. Nothing is more characteristic of the classical works of the great geometers of Greece, or more tantalising, than the absence of any indication of the steps by which they worked their way to the discovery of their great theorems. As they have come down to us, these theorems are finished masterpieces which leave no traces of any rough-hewn stage, no hint of the method by which they were evolved. We cannot but suppose that the

* *Hermes* XLII. 1907, pp. 243—297.
† *Bibliotheca Mathematica* VII₃, 1906–7, pp. 321—363.
‡ *Heronis Alexandrini opera*, Vol. III. 1903, pp. 80, 17; 130, 15; 130, 25.
§ Vide *The Works of Archimedes*, p. xxii.

INTRODUCTORY NOTE 7

Greeks had some method or methods of analysis hardly less powerful than those of modern analysis; yet, in general, they seem to have taken pains to clear away all traces of the machinery used and all the litter, so to speak, resulting from tentative efforts, before they permitted themselves to publish, in sequence carefully thought out, and with definitive and rigorously scientific proofs, the results obtained. A partial exception is now furnished by the *Method*; for here we have a sort of lifting of the veil, a glimpse of the interior of Archimedes' workshop as it were. He tells us how he discovered certain theorems in quadrature and cubature, and he is at the same time careful to insist on the difference between (1) the means which may be sufficient to suggest the truth of theorems, although not furnishing scientific proofs of them, and (2) the rigorous demonstrations of them by irrefragable geometrical methods which must follow before they can be finally accepted as established; to use Archimedes' own terms, the former enable theorems to be *investigated* ($\theta\epsilon\omega\rho\epsilon\hat{\iota}\nu$) but not to be *proved* ($\dot{\alpha}\pi o\delta\epsilon\iota\kappa\nu\acute{\nu}\nu\alpha\iota$). The mechanical method, then, used in our treatise and shown to be so useful for the discovery of theorems is distinctly said to be incapable of furnishing proofs of them; and Archimedes promises to add, as regards the two main theorems enunciated at the beginning, the necessary supplement in the shape of the formal geometrical proof. One of the two geometrical proofs is lost, but fragments of the other are contained in the MS. which are sufficient to show that the method was the orthodox method of exhaustion in the form in which Archimedes applies it elsewhere, and to enable the proof to be reconstructed.

The rest of this note will be best understood after the treatise itself has been read; but the essential features of the mechanical method employed by Archimedes are these. Suppose X to be a plane or solid figure, the area or content of which has to be found. The method is to weigh infinitesimal elements of X (with or without the addition of the corresponding elements of another figure C) against the corresponding elements of a figure B, B and C being such figures that their areas or volumes, and the position of the centre of gravity of B, are known beforehand. For this purpose the figures are first placed in such a position that they have, as common diameter or axis, one and the same straight line; if then the infinitesimal elements are sections of the figures made by parallel planes perpendicular (in general) to the axis and cutting the figures, the centres of gravity of all the elements lie at one point or other on the common diameter or axis. This diameter or axis is produced and is imagined to be the bar or lever of a balance. It is sufficient to take the simple case where the elements of X alone are weighed against the elements of another figure B. The elements which correspond to one another are the sections of X and B respectively by any one plane perpendicular (in general) to the diameter or axis and cutting both figures; the elements are spoken of as straight lines in the case of plane figures and as plane areas in the case of solid figures. Although Archimedes calls the elements straight lines and plane areas respectively, they are of course, in the first case, indefinitely narrow strips (areas) and, in the second case, indefinitely thin plane laminae (solids); but the breadth or thickness (dx, as we might call it) does not enter into the calculation because it is regarded as the same in each of the two corresponding elements which are separately weighed against each other, and therefore divides out. The number of the elements in each figure is infinite, but Archimedes has no need to say this; he merely says that X and B are *made up* of *all* the elements in them respectively, i.e. of the straight lines in the case of areas and of the plane areas in the case of solids.

The object of Archimedes is so to arrange the balancing of the elements that the elements of X are all applied at *one point* of the lever, while the elements of B operate at different points, namely where they actually are in the first instance. He contrives therefore to move the elements of X away from their first position and to concentrate them at one point on the lever, while the elements of B are left where they are, and so operate at their respective centres of gravity. Since the centre of gravity of B as a whole is known, as well as its area or volume, it may then be supposed to act as one mass applied at its centre of gravity; and consequently, taking the whole bodies X and B as ultimately placed respectively, we know the distances of the two centres of gravity from the fulcrum or point of suspension of the lever, and also the area or volume of B. Hence the area or volume of X is found. The method may be applied, conversely, to the problem of finding the centre of gravity of X when its area or volume is known beforehand; in this case it is necessary that the elements of X, and therefore X itself, should be weighed *in the places where they are*, and that the figures the elements of which are moved to one single point of the lever, to be weighed there, should be other figures and not X.

The method will be seen to be, not *integration*, as certain geometrical proofs in the great treatises actually are, but a clever device for *avoiding* the particular integration which would naturally be used to find directly the area or volume required, and making the solution depend, instead, upon *another* integration the result of which is already known. Archimedes deals with *moments* about the point of suspension of the lever, i.e. the products of the elements of area or volume into the distances between the point of suspension of the lever and the centres of gravity of the elements respectively; and, as we said above, while these distances are different for all the elements of B, he contrives, by moving the elements of X, to make them the same for all the elements of X in their final position. He assumes, as known, the fact that the sum of the moments of each particle of the figure B acting at the point where it is placed is equal to the moment of the whole figure applied as one mass at one point, its centre of gravity.

Suppose now that the element of X is $u\,.\,dx$, u being the length or area of a section of X by one of a whole series of parallel planes cutting the lever at right angles, x being measured along the lever (which is the common axis of the two figures) from the point of suspension of the lever as origin. This element is then supposed to be placed on the lever at a constant distance, say a, from the origin and on the opposite side of it from B. If $u'\,.\,dx$ is the corresponding element of B cut off by the same plane and x its distance from the origin, Archimedes' argument establishes the equation

$$a\int_h^k u\,dx = \int_h^k xu'\,dx.$$

Now the second integral is known because the area or volume of the figure B (say a triangle, a pyramid, a prism, a sphere, a cone, or a cylinder) is known, and it can be supposed to be applied as one mass at its centre of gravity, which is also known; the integral is equal to bU, where b is the distance of the centre of gravity from the point of suspension of the lever, and U is the area or content of B. Hence

$$\text{the area or volume of } X = \frac{bU}{a}.$$

In the case where the elements of X are weighed along with the corresponding elements of another figure C against corresponding elements of B, we have, if v be the element of C, and V its area or content,

$$a \int_h^k u\,dx + a \int_h^k v\,dx = \int_h^k x\,u'dx$$

and (area or volume of $X + V$) $a = bU$.

In the particular problems dealt with in the treatise h is always $= 0$, and k is often, but not always, equal to a.

Our admiration of the genius of the greatest mathematician of antiquity must surely be increased, if that were possible, by a perusal of the work before us. Mathematicians will doubtless agree that it is astounding that Archimedes, writing (say) about 250 B.C., should have been able to solve such problems as those of finding the volume and the centre of gravity of any segment of a sphere, and the centre of gravity of a semicircle, by a method so simple, a method too (be it observed) which would be quite rigorous enough for us to-day, although it did not satisfy Archimedes himself.

Apart from the mathematical content of the book, it is interesting, not only for Archimedes' explanations of the course which his investigations took, but also for the allusion to *Democritus* as the discoverer of the theorem that the volumes of a pyramid and a cone are one-third of the volumes of a prism and a cylinder respectively which have the same base and equal height. These propositions had always been supposed to be due to Eudoxus, and indeed Archimedes himself has a statement to this effect*. It now appears that, though Eudoxus was the first to prove them scientifically, Democritus was the first to assert their truth. I have elsewhere† made a suggestion as to the probable course of Democritus' argument, which, on Archimedes' view, did not amount to a proof of the propositions; but it may be well to re-state it here. Plutarch, in a well-known passage‡, speaks of Democritus as having raised the following question in natural philosophy ($\phi v \sigma \iota \kappa \hat{\omega} s$) : "if a cone were cut by a plane parallel to the base [by which is clearly meant a plane indefinitely near to the base], what must we think of the surfaces of the sections? Are they equal or unequal? For, if they are unequal, they will make the cone irregular, as having many indentations, like steps, and unevennesses; but, if they are equal, the sections will be equal, and the cone will appear to have the property of the cylinder and to be made up of equal, not unequal,

---

* *On the Sphere and Cylinder*, Preface to Book I.
† *The Thirteen Books of Euclid's Elements*, Vol. III. p. 368.
‡ Plutarch, *De Comm. Not. adv. Stoicos* XXXIX. 3.

circles, which is very absurd." The phrase "made up of equal... circles" (ἐξ ἴσων συγκείμενος...κύκλων) shows that Democritus already had the idea of a solid being the sum of an infinite number of parallel planes, or indefinitely thin laminae, indefinitely near together: a most important anticipation of the same thought which led to such fruitful results in Archimedes. If then we may make a conjecture as to Democritus' argument with regard to a pyramid, it seems probable that he would notice that, if two pyramids of the same height and with equal triangular bases are respectively cut by planes parallel to the base and dividing the heights in the same ratio, the corresponding sections of the two pyramids are equal, whence he would infer that the pyramids are equal because they are the sums of the same infinite numbers of equal plane sections or indefinitely thin laminae. (This would be a particular anticipation of Cavalieri's proposition that the areal or solid contents of two figures are equal if two sections of them taken at the same height, whatever the height may be, always give equal straight lines or equal surfaces respectively.) And Democritus would of course see that the three pyramids into which a prism on the same base and of equal height with the original pyramid is divided (as in Eucl. XII. 7) satisfy, in pairs, this test of equality, so that the pyramid would be one third part of the prism. The extension to a pyramid with a polygonal base would be easy. And Democritus may have stated the proposition for the cone (of course without an absolute proof) as a natural inference from the result of increasing indefinitely the number of sides in a regular polygon forming the base of a pyramid.

In accordance with the plan adopted in *The Works of Archimedes*, I have marked by inverted commas the passages which, on account of their importance, historically or otherwise, I have translated literally from the Greek; the rest of the tract is reproduced in modern notation and phraseology. Words and sentences in square brackets represent for the most part Heiberg's conjectural restoration (in his German translation) of what may be supposed to have been written in the places where the MS. is illegible; in a few cases where the gap is considerable a note in brackets indicates what the missing passage presumably contained and, so far as necessary, how the deficiency may be made good.

T. L. H.

7 *June* 1912.

THE METHOD OF ARCHIMEDES TREATING OF MECHANICAL PROBLEMS— TO ERATOSTHENES

"Archimedes to Eratosthenes greeting.

I sent you on a former occasion some of the theorems discovered by me, merely writing out the enunciations and inviting you to discover the proofs, which at the moment I did not give. The enunciations of the theorems which I sent were as follows.

1. If in a right prism with a parallelogrammic base a cylinder be inscribed which has its bases in the opposite parallelograms*, and its sides [i.e. four generators] on the remaining planes (faces) of the prism, and if through the centre of the circle which is the base of the cylinder and (through) one side of the square in the plane opposite to it a plane be drawn, the plane so drawn will cut off from the cylinder a segment which is bounded by two planes and the surface of the cylinder, one of the two planes being the plane which has been drawn and the other the plane in which the base of the cylinder is, and the surface being that which is between the said planes; and the segment cut off from the cylinder is one sixth part of the whole prism.

2. If in a cube a cylinder be inscribed which has its bases in the opposite parallelograms† and touches with its surface the remaining four planes (faces), and if there also be inscribed in the same cube another cylinder which has its bases in other parallelograms and touches with its surface the remaining four planes (faces), then the figure bounded by the surfaces of the cylinders, which is within both cylinders, is two-thirds of the whole cube.

Now these theorems differ in character from those communicated before; for we compared the figures then in question,

---
* The parallelograms are apparently *squares*. † i.e. squares.

conoids and spheroids and segments of them, in respect of size, with figures of cones and cylinders: but none of those figures have yet been found to be equal to a solid figure bounded by planes; whereas each of the present figures bounded by two planes and surfaces of cylinders is found to be equal to one of the solid figures which are bounded by planes. The proofs then of these theorems I have written in this book and now send to you. Seeing moreover in you, as I say, an earnest student, a man of considerable eminence in philosophy, and an admirer [of mathematical inquiry], I thought fit to write out for you and explain in detail in the same book the peculiarity of a certain method, by which it will be possible for you to get a start to enable you to investigate some of the problems in mathematics by means of mechanics. This procedure is, I am persuaded, no less useful even for the proof of the theorems themselves; for certain things first became clear to me by a mechanical method, although they had to be demonstrated by geometry afterwards because their investigation by the said method did not furnish an actual demonstration. But it is of course easier, when we have previously acquired, by the method, some knowledge of the questions, to supply the proof than it is to find it without any previous knowledge. This is a reason why, in the case of the theorems the proof of which Eudoxus was the first to discover, namely that the cone is a third part of the cylinder, and the pyramid of the prism, having the same base and equal height, we should give no small share of the credit to Democritus who was the first to make the assertion with regard to the said figure* though he did not prove it. I am myself in the position of having first made the discovery of the theorem now to be published [by the method indicated], and I deem it necessary to expound the method partly because I have already spoken of it† and I do not want to be thought to have uttered vain words, but

---

* περὶ τοῦ εἰρημένου σχήματος, in the singular. Possibly Archimedes may have thought of the case of the pyramid as being the more fundamental and as really involving that of the cone. Or perhaps "figure" may be intended for "type of figure."

† Cf. Preface to *Quadrature of Parabola*.

equally because I am persuaded that it will be of no little service to mathematics; for I apprehend that some, either of my contemporaries or of my successors, will, by means of the method when once established, be able to discover other theorems in addition, which have not yet occurred to me.

First then I will set out the very first theorem which became known to me by means of mechanics, namely that

Any segment of a section of a right-angled cone (i.e. a parabola) is four-thirds of the triangle which has the same base and equal height,

and after this I will give each of the other theorems investigated by the same method. Then, at the end of the book, I will give the geometrical [proofs of the propositions]...

[I premise the following propositions which I shall use in the course of the work.]

1. If from [one magnitude another magnitude be subtracted which has not the same centre of gravity, the centre of gravity of the remainder is found by] producing [the straight line joining the centres of gravity of the whole magnitude and of the subtracted part in the direction of the centre of gravity of the whole] and cutting off from it a length which has to the distance between the said centres of gravity the ratio which the weight of the subtracted magnitude has to the weight of the remainder.

[*On the Equilibrium of Planes*, I. 8]

2. If the centres of gravity of any number of magnitudes whatever be on the same straight line, the centre of gravity of the magnitude made up of all of them will be on the same straight line. [Cf. *Ibid.* I. 5]

3. The centre of gravity of any straight line is the point of bisection of the straight line. [Cf. *Ibid.* I. 4]

4. The centre of gravity of any triangle is the point in which the straight lines drawn from the angular points of the triangle to the middle points of the (opposite) sides cut one another. [*Ibid.* I. 13, 14]

5. The centre of gravity of any parallelogram is the point in which the diagonals meet. [*Ibid.* I. 10]

6. The centre of gravity of a circle is the point which is also the centre [of the circle].

7. The centre of gravity of any cylinder is the point of bisection of the axis.

8. The centre of gravity of any cone is [the point which divides its axis so that] the portion [adjacent to the vertex is] triple [of the portion adjacent to the base].

[All these propositions have already been] proved*. [Besides these I require also the following proposition, which is easily proved:

If in two series of magnitudes those of the first series are, in order, proportional to those of the second series and further] the magnitudes [of the first series], either all or some of them, are in any ratio whatever [to those of a third series], and if the magnitudes of the second series are in the same ratio to the corresponding magnitudes [of a fourth series], then the sum of the magnitudes of the first series has to the sum of the selected magnitudes of the third series the same ratio which the sum of the magnitudes of the second series has to the sum of the (correspondingly) selected magnitudes of the fourth series. [*On Conoids and Spheroids*, Prop. 1.]"

Proposition 1.

Let ABC be a segment of a parabola bounded by the straight line AC and the parabola ABC, and let D be the middle point of AC. Draw the straight line DBE parallel to the axis of the parabola and join AB, BC.

Then shall the segment ABC be $\frac{4}{3}$ of the triangle ABC.

From A draw AKF parallel to DE, and let the tangent to the parabola at C meet DBE in E and AKF in F. Produce CB to meet AF in K, and again produce CK to H, making KH equal to CK.

---

* The problem of finding the centre of gravity of a cone is not solved in any extant work of Archimedes. It may have been solved either in a separate treatise, such as the περὶ ζυγῶν, which is lost, or perhaps in a larger mechanical work of which the extant books *On the Equilibrium of Planes* formed only a part.

Consider CH as the bar of a balance, K being its middle point.

Let MO be any straight line parallel to ED, and let it meet CF, CK, AC in M, N, O and the curve in P.

Now, since CE is a tangent to the parabola and CD the semi-ordinate,
$$EB = BD;$$
"for this is proved in the Elements [of Conics]*."

Since FA, MO are parallel to ED, it follows that
$$FK = KA, \quad MN = NO.$$

Now, by the property of the parabola, "proved in a lemma,"
$$MO : OP = CA : AO \quad \text{[Cf. } Quadrature\ of\ Parabola,\ \text{Prop. 5]}$$
$$= CK : KN \quad \text{[Eucl. VI. 2]}$$
$$= HK : KN.$$

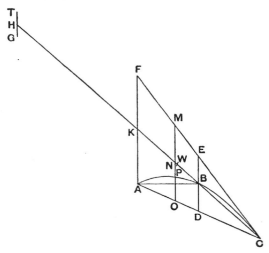

Take a straight line TG equal to OP, and place it with its centre of gravity at H, so that $TH = HG$; then, since N is the centre of gravity of the straight line MO,

and
$$MO : TG = HK : KN,$$

---

* i.e. the works on conics by Aristaeus and Euclid. Cf. the similar expression in *On Conoids and Spheroids*, Prop. 3, and *Quadrature of Parabola*, Prop. 3.

it follows that TG at H and MO at N will be in equilibrium about K. [*On the Equilibrium of Planes*, I. 6, 7]

Similarly, for all other straight lines parallel to DE and meeting the arc of the parabola, (1) the portion intercepted between FC, AC with its middle point on KC and (2) a length equal to the intercept between the curve and AC placed with its centre of gravity at H will be in equilibrium about K.

Therefore K is the centre of gravity of the whole system consisting (1) of all the straight lines as MO intercepted between FC, AC and placed as they actually are in the figure and (2) of all the straight lines placed at H equal to the straight lines as PO intercepted between the curve and AC.

And, since the triangle CFA is made up of all the parallel lines like MO,

and the segment CBA is made up of all the straight lines like PO within the curve,

it follows that the triangle, placed where it is in the figure, is in equilibrium about K with the segment CBA placed with its centre of gravity at H.

Divide KC at W so that $CK = 3KW$;

then W is the centre of gravity of the triangle ACF; " for this is proved in the books on equilibrium " (ἐν τοῖς ἰσορροπικοῖς).

[Cf. *On the Equilibrium of Planes* I. 15]

Therefore $\triangle ACF$: (segment ABC) = HK : KW

$$= 3 : 1.$$

Therefore segment $ABC = \frac{1}{3} \triangle ACF$.

But $\triangle ACF = 4 \triangle ABC$.

Therefore segment $ABC = \frac{4}{3} \triangle ABC$.

"Now the fact here stated is not actually demonstrated by the argument used; but that argument has given a sort of indication that the conclusion is true. Seeing then that the theorem is not demonstrated, but at the same time suspecting that the conclusion is true, we shall have recourse to the geometrical demonstration which I myself discovered and have already published*."

Proposition 2.

We can investigate by the same method the propositions that

(1) *Any sphere is (in respect of solid content) four times the cone with base equal to a great circle of the sphere and height equal to its radius; and*

(2) *the cylinder with base equal to a great circle of the sphere and height equal to the diameter is* $1\frac{1}{2}$ *times the sphere.*

(1) Let $ABCD$ be a great circle of a sphere, and AC, BD diameters at right angles to one another.

Let a circle be drawn about BD as diameter and in a plane perpendicular to AC, and on this circle as base let a cone be described with A as vertex. Let the surface of this cone be produced and then cut by a plane through C parallel to its base; the section will be a circle on EF as diameter. On this circle as base let a cylinder be erected with height and axis AC, and produce CA to H, making AH equal to CA.

Let CH be regarded as the bar of a balance, A being its middle point.

Draw any straight line MN in the plane of the circle $ABCD$ and parallel to BD. Let MN meet the circle in O, P, the diameter AC in S, and the straight lines AE, AF in Q, R respectively. Join AO.

---

* The word governing τὴν γεωμετρουμένην ἀπόδειξιν in the Greek text is τάξομεν, a reading which seems to be doubtful and is certainly difficult to translate. Heiberg translates as if τάξομεν meant "we shall give lower down" or "later on," but I agree with Th. Reinach (*Revue générale des sciences pures et appliquées*, 30 November 1907, p. 918) that it is questionable whether Archimedes would really have written out in full once more, as an appendix, a proof which, as he says, had already been published (i.e. presumably in the *Quadrature of a Parabola*). τάξομεν, if correct, should apparently mean "we shall appoint," "prescribe" or "assign."

Through MN draw a plane at right angles to AC; this plane will cut the cylinder in a circle with diameter MN, the sphere in a circle with diameter OP, and the cone in a circle with diameter QR.

Now, since $MS = AC$, and $QS = AS$,
$$MS \cdot SQ = CA \cdot AS$$
$$= AO^2$$
$$= OS^2 + SQ^2.$$

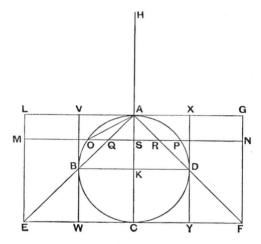

And, since $HA = AC$,
$$HA : AS = CA : AS$$
$$= MS : SQ$$
$$= MS^2 : MS \cdot SQ$$
$$= MS^2 : (OS^2 + SQ^2), \text{ from above,}$$
$$= MN^2 : (OP^2 + QR^2)$$
$$= (\text{circle, diam. } MN) : (\text{circle, diam. } OP$$
$$+ \text{ circle, diam. } QR).$$

That is,

$HA : AS = (\text{circle in cylinder}) : (\text{circle in sphere} + \text{circle in cone})$.

Therefore the circle in the cylinder, placed where it is, is in equilibrium, about A, with the circle in the sphere together with the circle in the cone, if both the latter circles are placed with their centres of gravity at H.

Similarly for the three corresponding sections made by a plane perpendicular to AC and passing through any other straight line in the parallelogram LF parallel to EF.

If we deal in the same way with all the sets of three circles in which planes perpendicular to AC cut the cylinder, the sphere and the cone, and which make up those solids respectively, it follows that the cylinder, in the place where it is, will be in equilibrium about A with the sphere and the cone together, when both are placed with their centres of gravity at H.

Therefore, since K is the centre of gravity of the cylinder,

$HA : AK =$ (cylinder) : (sphere + cone AEF).

But $HA = 2AK$;
therefore   cylinder = 2 (sphere + cone AEF).

Now   cylinder = 3 (cone AEF);   [Eucl. XII. 10]
therefore   cone AEF = 2 (sphere).

But, since $EF = 2BD$,
        cone AEF = 8 (cone ABD);
therefore   sphere = 4 (cone ABD).

(2) Through B, D draw VBW, XDY parallel to AC; and imagine a cylinder which has AC for axis and the circles on VX, WY as diameters for bases.

Then   cylinder $VY = 2$ (cylinder VD)
              $= 6$ (cone ABD)   [Eucl. XII. 10]
              $= \frac{3}{2}$ (sphere), from above.

Q.E.D.

"From this theorem, to the effect that a sphere is four times as great as the cone with a great circle of the sphere as base and with height equal to the radius of the sphere, I conceived the notion that the surface of any sphere is four times as great as a great circle in it; for, judging from the fact that any circle is equal to a triangle with base equal to the circumference and height equal to the radius of the circle, I apprehended that, in like manner, any sphere is equal to a cone with base equal to the surface of the sphere and height equal to the radius*."

Proposition 3.

By this method we can also investigate the theorem that

A cylinder with base equal to the greatest circle in a spheroid and height equal to the axis of the spheroid is $1\frac{1}{2}$ times the spheroid;

and, when this is established, it is plain that

If any spheroid be cut by a plane through the centre and at right angles to the axis, the half of the spheroid is double of the cone which has the same base and the same axis as the segment (i.e. the half of the spheroid).

Let a plane through the axis of a spheroid cut its surface in the ellipse $ABCD$, the diameters (i.e. axes) of which are AC, BD; and let K be the centre.

Draw a circle about BD as diameter and in a plane perpendicular to AC;

imagine a cone with this circle as base and A as vertex produced and cut by a plane through C parallel to its base; the section will be a circle in a plane at right angles to AC and about EF as diameter.

Imagine a cylinder with the latter circle as base and axis AC; produce CA to H, making AH equal to CA.

Let HC be regarded as the bar of a balance, A being its middle point.

In the parallelogram LF draw any straight line MN parallel to EF meeting the ellipse in O, P and AE, AF, AC in Q, R, S respectively.

---

* That is to say, Archimedes originally solved the problem of finding the solid content of a sphere before that of finding its surface, and he inferred the result of the latter problem from that of the former. Yet in *On the Sphere and Cylinder* I. the surface is independently found (Prop. 33) and *before* the volume, which is found in Prop. 34 : another illustration of the fact that the order of propositions in the treatises of the Greek geometers as finally elaborated does not necessarily follow the order of discovery.

If now a plane be drawn through MN at right angles to AC, it will cut the cylinder in a circle with diameter MN, the spheroid in a circle with diameter OP, and the cone in a circle with diameter QR.

Since $HA = AC$,
$$HA : AS = CA : AS$$
$$= EA : AQ$$
$$= MS : SQ.$$

Therefore $HA : AS = MS^2 : MS \cdot SQ.$

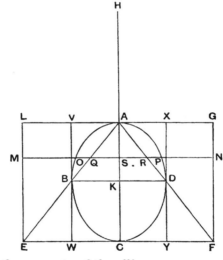

But, by the property of the ellipse,
$$AS \cdot SC : SO^2 = AK^2 : KB^2$$
$$= AS^2 : SQ^2;$$
therefore $SQ^2 : SO^2 = AS^2 : AS \cdot SC$
$$= SQ^2 : SQ \cdot QM,$$
and accordingly $SO^2 = SQ \cdot QM.$

Add SQ^2 to each side, and we have
$$SO^2 + SQ^2 = SQ \cdot SM.$$

Therefore, from above, we have
$HA : AS = MS^2 : (SO^2 + SQ^2)$
$$= MN^2 : (OP^2 + QR^2)$$
$$= (\text{circle, diam. } MN) : (\text{circle, diam. } OP + \text{circle, diam. } QR)$$

That is,

$HA : AS =$ (circle in cylinder) : (circle in spheroid + circle in cone).

Therefore the circle in the cylinder, in the place where it is, is in equilibrium, about A, with the circle in the spheroid and the circle in the cone together, if both the latter circles are placed with their centres of gravity at H.

Similarly for the three corresponding sections made by a plane perpendicular to AC and passing through any other straight line in the parallelogram LF parallel to EF.

If we deal in the same way with all the sets of three circles in which planes perpendicular to AC cut the cylinder, the spheroid and the cone, and which make up those figures respectively, it follows that the cylinder, in the place where it is, will be in equilibrium about A with the spheroid and the cone together, when both are placed with their centres of gravity at H.

Therefore, since K is the centre of gravity of the cylinder,

$HA : AK =$ (cylinder) : (spheroid + cone AEF).

But $HA = 2AK$;

therefore   cylinder $= 2$ (spheroid + cone AEF).

And   cylinder $= 3$ (cone AEF);   [Eucl. XII. 10]
therefore   cone $AEF = 2$ (spheroid).

But, since $EF = 2BD$,

cone $AEF = 8$ (cone ABD);

therefore   spheroid $= 4$ (cone ABD),
and   half the spheroid $= 2$ (cone ABD).

Through B, D draw VBW, XDY parallel to AC; and imagine a cylinder which has AC for axis and the circles on VX, WY as diameters for bases.

Then   cylinder $VY = 2$ (cylinder VD)
               $= 6$ (cone ABD)
               $= \frac{3}{2}$ (spheroid), from above.

Q.E.D.

Proposition 4.

Any segment of a right-angled conoid (i.e. a paraboloid of revolution) cut off by a plane at right angles to the axis is $1\frac{1}{2}$ times the cone which has the same base and the same axis as the segment.

This can be investigated by our method, as follows.

Let a paraboloid of revolution be cut by a plane through the axis in the parabola BAC;

and let it also be cut by another plane at right angles to the axis and intersecting the former plane in BC. Produce DA, the axis of the segment, to H, making HA equal to AD.

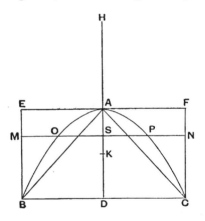

Imagine that HD is the bar of a balance, A being its middle point.

The base of the segment being the circle on BC as diameter and in a plane perpendicular to AD,

imagine (1) a cone drawn with the latter circle as base and A as vertex, and (2) a cylinder with the same circle as base and AD as axis.

In the parallelogram EC let any straight line MN be drawn parallel to BC, and through MN let a plane be drawn at right angles to AD; this plane will cut the cylinder in a circle with diameter MN and the paraboloid in a circle with diameter OP.

THE METHOD

Now, BAC being a parabola and BD, OS ordinates,
$$DA : AS = BD^2 : OS^2,$$
or $$HA : AS = MS^2 : SO^2.$$
Therefore
$HA : AS =$ (circle, rad. MS) : (circle, rad. OS)
$=$ (circle in cylinder) : (circle in paraboloid).

Therefore the circle in the cylinder, in the place where it is, will be in equilibrium, about A, with the circle in the paraboloid, if the latter is placed with its centre of gravity at H.

Similarly for the two corresponding circular sections made by a plane perpendicular to AD and passing through any other straight line in the parallelogram which is parallel to BC.

Therefore, as usual, if we take all the circles making up the whole cylinder and the whole segment and treat them in the same way, we find that the cylinder, in the place where it is, is in equilibrium about A with the segment placed with its centre of gravity at H.

If K is the middle point of AD, K is the centre of gravity of the cylinder;

therefore $HA : AK =$ (cylinder) : (segment).
Therefore cylinder $= 2$ (segment).
And cylinder $= 3$ (cone ABC); [Eucl. XII. 10]
therefore segment $= \frac{3}{2}$ (cone ABC).

Proposition 5.

The centre of gravity of a segment of a right-angled conoid (i.e. a paraboloid of revolution) cut off by a plane at right angles to the axis is on the straight line which is the axis of the segment, and divides the said straight line in such a way that the portion of it adjacent to the vertex is double of the remaining portion.

This can be investigated by the method, as follows.

Let a paraboloid of revolution be cut by a plane through the axis in the parabola BAC;

and let it also be cut by another plane at right angles to the axis and intersecting the former plane in BC.

Produce DA, the axis of the segment, to H, making HA equal to AD; and imagine DH to be the bar of a balance, its middle point being A.

The base of the segment being the circle on BC as diameter and in a plane perpendicular to AD, imagine a cone with this circle as base and A as vertex, so that AB, AC are generators of the cone.

In the parabola let any double ordinate OP be drawn meeting AB, AD, AC in Q, S, R respectively.

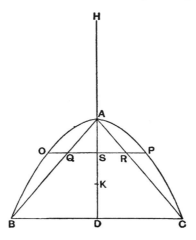

Now, from the property of the parabola,
$$BD^2 : OS^2 = DA : AS$$
$$= BD : QS$$
$$= BD^2 : BD . QS.$$

Therefore $OS^2 = BD . QS,$

or $BD : OS = OS : QS,$

whence $BD : QS = OS^2 : QS^2.$

But $BD : QS = AD : AS$
$$= HA : AS.$$

Therefore $HA : AS = OS^2 : QS^2$
$$= OP^2 : QR^2.$$

If now through OP a plane be drawn at right angles to AD, this plane cuts the paraboloid in a circle with diameter OP and the cone in a circle with diameter QR.

We see therefore that

$HA : AS =$ (circle, diam. OP) : (circle, diam. QR)
$=$ (circle in paraboloid) : (circle in cone);

and the circle in the paraboloid, in the place where it is, is in equilibrium about A with the circle in the cone placed with its centre of gravity at H.

Similarly for the two corresponding circular sections made by a plane perpendicular to AD and passing through any other ordinate of the parabola.

Dealing therefore in the same way with all the circular sections which make up the whole of the segment of the paraboloid and the cone respectively, we see that the segment of the paraboloid, in the place where it is, is in equilibrium about A with the cone placed with its centre of gravity at H.

Now, since A is the centre of gravity of the whole system as placed, and the centre of gravity of part of it, namely the cone, as placed, is at H, the centre of gravity of the rest, namely the segment, is at a point K on HA produced such that

$HA : AK =$ (segment) : (cone).

But segment $= \frac{3}{2}$ (cone). [Prop. 4]

Therefore $HA = \frac{3}{2} AK$;

that is, K divides AD in such a way that $AK = 2KD$.

Proposition 6.

The centre of gravity of any hemisphere [is on the straight line which] is its axis, and divides the said straight line in such a way that the portion of it adjacent to the surface of the hemisphere has to the remaining portion the ratio which 5 has to 3.

Let a sphere be cut by a plane through its centre in the circle $ABCD$;

let AC, BD be perpendicular diameters of this circle, and through BD let a plane be drawn at right angles to AC.

The latter plane will cut the sphere in a circle on BD as diameter.

Imagine a cone with the latter circle as base and A as vertex.

Produce CA to H, making AH equal to CA, and let HC be regarded as the bar of a balance, A being its middle point.

In the semicircle BAD, let any straight line OP be drawn parallel to BD and cutting AC in E and the two generators AB, AD of the cone in Q, R respectively. Join AO.

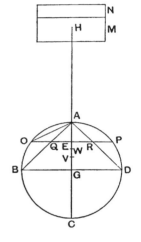

Through OP let a plane be drawn at right angles to AC; this plane will cut the hemisphere in a circle with diameter OP and the cone in a circle with diameter QR.

Now
$$HA:AE = AC:AE$$
$$= AO^2 : AE^2$$
$$= (OE^2 + AE^2) : AE^2$$
$$= (OE^2 + QE^2) : QE^2$$
$$= (\text{circle, diam. } OP + \text{circle, diam. } QR) : (\text{circle, diam. } QR).$$

Therefore the circles with diameters OP, QR, in the places where they are, are in equilibrium about A with the circle with diameter QR if the latter is placed with its centre of gravity at H.

And, since the centre of gravity of the two circles with diameters OP, QR taken together, in the place where they are, is......

[There is a lacuna here; but the proof can easily be completed on the lines of the corresponding but more difficult case in Prop. 8.

We proceed thus from the point where the circles with diameters OP, QR, in the place where they are, balance, about A, the circle with diameter QR placed with its centre of gravity at H.

A similar relation holds for all the other sets of circular sections made by other planes passing through points on AG and at right angles to AG.

Taking then all the circles which fill up the hemisphere BAD and the cone ABD respectively, we find that
the hemisphere BAD and the cone ABD, in the places where they are, together balance, about A, a cone equal to ABD placed with its centre of gravity at H.

Let the cylinder $M+N$ be equal to the cone ABD.

Then, since the cylinder $M+N$ placed with its centre of gravity at H balances the hemisphere BAD and the cone ABD in the places where they are,
suppose that the portion M of the cylinder, placed with its centre of gravity at H, balances the cone ABD (alone) in the place where it is; therefore the portion N of the cylinder placed with its centre of gravity at H balances the hemisphere (alone) in the place where it is.

Now the centre of gravity of the cone is at a point V such that $AG = 4GV$;
therefore, since M at H is in equilibrium with the cone,
$$M : (\text{cone}) = \tfrac{3}{4}AG : HA = \tfrac{3}{8}AC : AC,$$
whence $\qquad M = \tfrac{3}{8}(\text{cone}).$

But $M + N = (\text{cone})$; therefore $N = \tfrac{5}{8}(\text{cone})$.

Now let the centre of gravity of the hemisphere be at W, which is somewhere on AG.

Then, since N at H balances the hemisphere alone,
$$(\text{hemisphere}) : N = HA : AW.$$
But the hemisphere $BAD =$ twice the cone ABD;
[*On the Sphere and Cylinder* I. 34 and Prop. 2 above]
and $N = \tfrac{5}{8}(\text{cone})$, from above.

Therefore $\qquad 2 : \tfrac{5}{8} = HA : AW$
$$= 2AG : AW,$$
whence $AW = \tfrac{5}{8}AG$, so that W divides AG in such a way that
$$AW : WG = 5 : 3.]$$

Proposition 7.

We can also investigate by the same method the theorem that

[*Any segment of a sphere has*] *to the cone* [*with the same base and height the ratio which the sum of the radius of the sphere and the height of the complementary segment has to the height of the complementary segment.*]

[There is a lacuna here; but all that is missing is the construction, and the construction is easily understood by

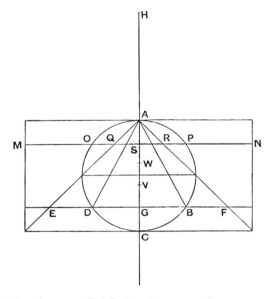

means of the figure. *BAD* is of course the segment of the sphere the volume of which is to be compared with the volume of a cone with the same base and height.]

The plane drawn through *MN* and at right angles to *AC* will cut the cylinder in a circle with diameter *MN*, the segment of the sphere in a circle with diameter *OP*, and the cone on the base *EF* in a circle with diameter *QR*.

In the same way as before [cf. Prop. 2] we can prove that the circle with diameter *MN*, in the place where it is, is in equilibrium about A with the two circles with diameters OP, QR if these circles are both moved and placed with their centres of gravity at H.

The same thing can be proved of all sets of three circles in which the cylinder, the segment of the sphere, and the cone with the common height AG are all cut by any plane perpendicular to AC.

Since then the sets of circles make up the whole cylinder, the whole segment of the sphere and the whole cone respectively, it follows that the cylinder, in the place where it is, is in equilibrium about A with the sum of the segment of the sphere and the cone if both are placed with their centres of gravity at H.

Divide AG at W, V in such a way that
$$AW = WG, \quad AV = 3VG.$$

Therefore W will be the centre of gravity of the cylinder, and V will be the centre of gravity of the cone.

Since, now, the bodies are in equilibrium as described,

(cylinder) : (cone AEF + segment BAD of sphere)
$= HA : AW$.

....

[The rest of the proof is lost; but it can easily be supplied thus.

We have

(cone AEF + segmt. BAD) : (cylinder) $= AW : AC$
$= AW . AC : AC^2$.

But (cylinder) : (cone AEF) $= AC^2 : \frac{1}{3}EG^2$
$= AC^2 : \frac{1}{3}AG^2$.

Therefore, *ex aequali*,

(cone AEF + segmt. BAD) : (cone AEF) $= AW . AC : \frac{1}{3}AG^2$
$= \frac{1}{2}AC : \frac{1}{3}AG$,

whence (segmt. BAD) : (cone AEF) $= (\frac{1}{2}AC - \frac{1}{3}AG) : \frac{1}{3}AG$.

Again (cone AEF) : (cone ABD) $= EG^2 : DG^2$
$= AG^2 : AG . GC$
$= AG : GC$
$= \frac{1}{3}AG : \frac{1}{3}GC$.

Therefore, *ex aequali*,

(segment BAD) : (cone ABD) $= (\tfrac{1}{2}AC - \tfrac{1}{3}AG) : \tfrac{1}{3}GC$
$= (\tfrac{3}{2}AC - AG) : GC$
$= (\tfrac{1}{2}AC + GC) : GC.$

Q.E.D.]

Proposition 8.

[The enunciation, the setting-out, and a few words of the construction are missing.

The enunciation however can be supplied from that of Prop. 9, with which it must be identical except that it cannot refer to "*any* segment," and the presumption therefore is that the proposition was enunciated with reference to one kind of segment only, i.e. either a segment greater than a hemisphere or a segment less than a hemisphere.

Heiberg's figure corresponds to the case of a segment greater than a hemisphere. The segment investigated is of course the segment BAD. The setting-out and construction are self-evident from the figure.]

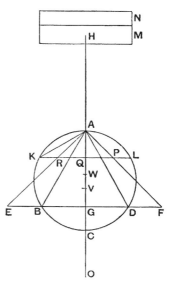

Produce AC to H, O, making HA equal to AC and CO equal to the radius of the sphere;

and let HC be regarded as the bar of a balance, the middle point being A.

In the plane cutting off the segment describe a circle with G as centre and radius (GE) equal to AG; and on this circle as base, and with A as vertex, let a cone be described. AE, AF are generators of this cone.

Draw KL, through any point Q on AG, parallel to EF and cutting the segment in K, L, and AE, AF in R, P respectively. Join AK.

THE METHOD

Now $HA : AQ = CA : AQ$
$$= AK^2 : AQ^2$$
$$= (KQ^2 + QA^2) : QA^2$$
$$= (KQ^2 + PQ^2) : PQ^2$$
$$= (\text{circle, diam. } KL + \text{circle, diam. } PR)$$
$$: (\text{circle, diam. } PR).$$

Imagine a circle equal to the circle with diameter PR placed with its centre of gravity at H;

therefore the circles on diameters KL, PR, in the places where they are, are in equilibrium about A with the circle with diameter PR placed with its centre of gravity at H.

Similarly for the corresponding circular sections made by any other plane perpendicular to AG.

Therefore, taking all the circular sections which make up the segment ABD of the sphere and the cone AEF respectively, we find that the segment ABD of the sphere and the cone AEF, in the places where they are, are in equilibrium with the cone AEF assumed to be placed with its centre of gravity at H.

Let the cylinder $M + N$ be equal to the cone AEF which has A for vertex and the circle on EF as diameter for base.

Divide AG at V so that $AG = 4VG$;

therefore V is the centre of gravity of the cone AEF; "for this has been proved before*."

Let the cylinder $M + N$ be cut by a plane perpendicular to the axis in such a way that the cylinder M (alone), placed with its centre of gravity at H, is in equilibrium with the cone AEF.

Since $M + N$ suspended at H is in equilibrium with the segment ABD of the sphere and the cone AEF in the places where they are,

while M, also at H, is in equilibrium with the cone AEF in the place where it is, it follows that

N at H is in equilibrium with the segment ABD of the sphere in the place where it is.

* Cf. note on p. 15 above.

Now (segment ABD of sphere) : (cone ABD)
$$= OG : GC;$$
"for this is already proved" [Cf. *On the Sphere and Cylinder* II. 2 Cor. as well as Prop. 7 *ante*].

And (cone ABD) : (cone AEF)
$$= \text{(circle, diam. } BD\text{)} : \text{(circle, diam. } EF\text{)}$$
$$= BD^2 : EF^2$$
$$= BG^2 : GE^2$$
$$= CG \cdot GA : GA^2$$
$$= CG : GA.$$

Therefore, *ex aequali*,
(segment ABD of sphere) : (cone AEF)
$$= OG : GA.$$

Take a point W on AG such that
$$AW : WG = (GA + 4GC) : (GA + 2GC).$$
We have then, inversely,
$$GW : WA = (2GC + GA) : (4GC + GA),$$
and, *componendo*,
$$GA : AW = (6GC + 2GA) : (4GC + GA).$$

But $GO = \frac{1}{4}(6GC + 2GA)$, [for $GO - GC = \frac{1}{2}(CG + GA)$]
and $CV = \frac{1}{4}(4GC + GA);$
therefore $GA : AW = OG : CV,$
and, alternately and inversely,
$$OG : GA = CV : WA.$$

It follows, from above, that
(segment ABD of sphere) : (cone AEF) $= CV : WA$.

Now, since the cylinder M with its centre of gravity at H is in equilibrium about A with the cone AEF with its centre of gravity at V,
$$\text{(cone } AEF\text{)} : \text{(cylinder } M\text{)} = HA : AV$$
$$= CA : AV;$$
and, since the cone AEF = the cylinder $M + N$, we have, *dividendo* and *invertendo*,
(cylinder M) : (cylinder N) $= AV : CV$.

Hence, *componendo*,

(cone AEF) : (cylinder N) = $CA : CV$*
$= HA : CV$.

But it was proved that

(segment ABD of sphere) : (cone AEF) = $CV : WA$;

therefore, *ex aequali*,

(segment ABD of sphere) : (cylinder N) = $HA : AW$.

And it was above proved that the cylinder N at H is in equilibrium about A with the segment ABD, in the place where it is;

therefore, since H is the centre of gravity of the cylinder N, W is the centre of gravity of the segment ABD of the sphere.

Proposition 9.

In the same way we can investigate the theorem that

The centre of gravity of any segment of a sphere is on the straight line which is the axis of the segment, and divides this straight line in such a way that the part of it adjacent to the vertex of the segment has to the remaining part the ratio which the sum of the axis of the segment and four times the axis of the complementary segment has to the sum of the axis of the segment and double the axis of the complementary segment.

[As this theorem relates to "*any* segment" but states the same result as that proved in the preceding proposition, it follows that Prop. 8 must have related to one kind of segment, either a segment greater than a semicircle (as in Heiberg's figure of Prop. 8) or a segment less than a semicircle; and the present proposition completed the proof for both kinds of segments. It would only require a slight change in the figure, in any case.]

Proposition 10.

By this method too we can investigate the theorem that

[*A segment of an obtuse-angled conoid (i.e. a hyperboloid of revolution) has to the cone which has*] *the same base* [*as the*

---

* Archimedes arrives at this result in a very roundabout way, seeing that it could have been obtained at once *convertendo*. Cf. Euclid x. 14.

segment and equal height the same ratio as the sum of the axis of the segment and three times] *the* "*annex to the axis*" (*i.e. half the transverse axis of the hyperbolic section through the axis of the hyperboloid or, in other words, the distance between the vertex of the segment and the vertex of the enveloping cone*) *has to the sum of the axis of the segment and double of the* "*annex*" *
[this is the theorem proved in *On Conoids and Spheroids*, Prop. 25], "and also many other theorems, which, as the method has been made clear by means of the foregoing examples, I will omit, in order that I may now proceed to compass the proofs of the theorems mentioned above."

Proposition 11.

If in a right prism with square bases a cylinder be inscribed having its bases in opposite square faces and touching with its surface the remaining four parallelogrammic faces, and if through the centre of the circle which is the base of the cylinder and one side of the opposite square face a plane be drawn, the figure cut off by the plane so drawn is one sixth part of the whole prism.

"This can be investigated by the method, and, when it is set out, I will go back to the proof of it by geometrical considerations."

[The investigation by the mechanical method is contained in the two Propositions, 11, 12. Prop. 13 gives another solution which, although it contains no mechanics, is still of the character which Archimedes regards as inconclusive, since it assumes that the solid is actually *made up* of parallel plane sections and that an auxiliary parabola is actually *made up* of parallel straight lines in it. Prop. 14 added the conclusive geometrical proof.]

Let there be a right prism with a cylinder inscribed as stated.

---

* The text has "triple" ($\tau\rho\iota\pi\lambda\alpha\sigma\iota\alpha\nu$) in the last line instead of "double." As there is a considerable lacuna before the last few lines, a theorem about the centre of gravity of a segment of a hyperboloid of revolution may have fallen out.

THE METHOD

Let the prism be cut through the axis of the prism and cylinder by a plane perpendicular to the plane which cuts off the portion of the cylinder; let this plane make, as section, the parallelogram AB, and let it cut the plane cutting off the portion of the cylinder (which plane is perpendicular to AB) in the straight line BC.

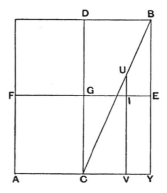

Let CD be the axis of the prism and cylinder, let EF bisect it at right angles, and through EF let a plane be drawn at right angles to CD; this plane will cut the prism in a square and the cylinder in a circle.

Let MN be the square and $OPQR$ the circle, and let the circle touch the sides of the square in O, P, Q, R [F, E in the first figure are identical with O, Q respectively]. Let H be the centre of the circle.

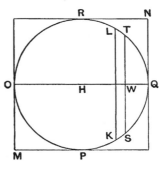

Let KL be the intersection of the plane through EF perpendicular to the axis of the cylinder and the plane cutting off the portion of the cylinder; KL is bisected by OHQ [and passes through the middle point of HQ].

Let any chord of the circle, as ST, be drawn perpendicular to HQ, meeting HQ in W; and through ST let a plane be drawn at right angles to OQ and produced on both sides of the plane of the circle $OPQR$.

The plane so drawn will cut the half cylinder having the semicircle PQR for section and the axis of the prism for height in a parallelogram, one side of which is equal to ST and another is a generator of the cylinder; and it will also cut the portion of the cylinder cut off in a parallelogram, one side of which is equal to ST and the other is equal and parallel to UV (in the first figure).

UV will be parallel to BY and will cut off, along EG in the parallelogram DE, the segment EI equal to QW.

Now, since EC is a parallelogram, and VI is parallel to GC,
$$EG : GI = YC : CV$$
$$= BY : UV$$
$$= (\square \text{ in half cyl.}) : (\square \text{ in portion of cyl.}).$$

And $EG = HQ$, $GI = HW$, $QH = OH$;
therefore $OH : HW = (\square \text{ in half cyl.}) : (\square \text{ in portion})$.

Imagine that the parallelogram in the portion of the cylinder is moved and placed at O so that O is its centre of gravity, and that OQ is the bar of a balance, H being its middle point.

Then, since W is the centre of gravity of the parallelogram in the half cylinder, it follows from the above that the parallelogram in the half cylinder, in the place where it is, with its centre of gravity at W, is in equilibrium about H with the parallelogram in the portion of the cylinder when placed with its centre of gravity at O.

Similarly for the other parallelogrammic sections made by any plane perpendicular to OQ and passing through any other chord in the semicircle PQR perpendicular to OQ.

If then we take all the parallelograms making up the half cylinder and the portion of the cylinder respectively, it follows that the half cylinder, in the place where it is, is in equilibrium about H with the portion of the cylinder cut off when the latter is placed with its centre of gravity at O.

Proposition 12.

Let the parallelogram (square) MN perpendicular to the axis, with the circle $OPQR$ and its diameters OQ, PR, be drawn separately.

THE METHOD

Join HG, HM, and through them draw planes at right angles to the plane of the circle, producing them on both sides of that plane.

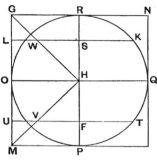

This produces a prism with triangular section GHM and height equal to the axis of the cylinder; this prism is $\frac{1}{4}$ of the original prism circumscribing the cylinder.

Let LK, UT be drawn parallel to OQ and equidistant from it, cutting the circle in K, T, RP in S, F, and GH, HM in W, V respectively.

Through LK, UT draw planes at right angles to PR, producing them on both sides of the plane of the circle; these planes produce as sections in the half cylinder PQR and in the prism GHM four parallelograms in which the heights are equal to the axis of the cylinder, and the other sides are equal to KS, TF, LW, UV respectively............

....

[The rest of the proof is missing, but, as Zeuthen says*, the result obtained and the method of arriving at it are plainly indicated by the above.

Archimedes wishes to prove that the half cylinder PQR, in the place where it is, balances the prism GHM, in the place where it is, about H as fixed point.

He has first to prove that the elements (1) the parallelogram with side $= KS$ and (2) the parallelogram with side $= LW$, in the places where they are, balance about S, or, in other words that the straight lines SK, LW, in the places where they are, balance about S.

Now (radius of circle $OPQR)^2 = SK^2 + SH^2$,
or $SL^2 = SK^2 + SW^2$.
Therefore $LS^2 - SW^2 = SK^2$,
and accordingly $(LS + SW) \cdot LW = SK^2$,
whence $\frac{1}{2}(LS + SW) : \frac{1}{2} SK = SK : LW$.

* Zeuthen in *Bibliotheca Mathematica* VII$_3$, 1906-7, pp. 352-3.

And $\frac{1}{2}(LS+SW)$ is the distance of the centre of gravity of LW from S,
while $\frac{1}{2}SK$ is the distance of the centre of gravity of SK from S.

Therefore SK and LW, in the places where they are, balance about S.

Similarly for the corresponding parallelograms.

Taking *all* the parallelogrammic elements in the half cylinder and prism respectively, we find that
the half cylinder PQR and the prism GHM, in the places where they are respectively, balance about H.

From this result and that of Prop. 11 we can at once deduce the volume of the portion cut off from the cylinder. For in Prop. 11 the portion of the cylinder, placed with its centre of gravity at O, is shown to balance (about H) the half-cylinder in the place where it is. By Prop. 12 we may substitute for the half-cylinder in the place where it is the prism GHM of that proposition turned the opposite way relatively to RP. The centre of gravity of the prism as thus placed is at a point (say Z) on HQ such that $HZ = \frac{2}{3}HQ$.

Therefore, assuming the prism to be applied at its centre of gravity, we have

(portion of cylinder) : (prism) $= \frac{2}{3}HQ : OH$
$= 2 : 3$;

therefore (portion of cylinder) $= \frac{2}{3}$ (prism GHM)
$= \frac{1}{6}$ (original prism).

Note. This proposition of course solves the problem of finding the centre of gravity of a half cylinder or, in other words, of a semicircle.

For the triangle GHM in the place where it is balances, about H, the semicircle PQR in the place where it is.

If then X is the point on HQ which is the centre of gravity of the semicircle,

$\frac{2}{3}HO \cdot (\triangle GHM) = HX \cdot (\text{semicircle } PQR)$,

or $\qquad \frac{2}{3}HO \cdot HO^2 = HX \cdot \frac{1}{2}\pi \cdot HO^2$;

that is, $\qquad HX = \dfrac{4}{3\pi} \cdot HQ.$]

Proposition 13.

Let there be a right prism with square bases, one of which is $ABCD$; in the prism let a cylinder be inscribed, the base of which is the circle $EFGH$ touching the sides of the square $ABCD$ in E, F, G, H.

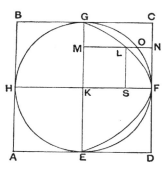

Through the centre and through the side corresponding to CD in the square face *opposite* to $ABCD$ let a plane be drawn; this will cut off a prism equal to $\frac{1}{4}$ of the original prism and formed by three parallelograms and two triangles, the triangles forming opposite faces.

In the semicircle EFG describe the parabola which has FK for axis and passes through E, G; draw MN parallel to KF meeting GE in M, the parabola in L, the semicircle in O and CD in N.

Then $MN \cdot NL = NF^2$;

"for this is clear." [Cf. Apollonius, *Conics* I. 11]
[The parameter is of course equal to GK or KF.]

Therefore $MN : NL = GK^2 : LS^2$.

Through MN draw a plane at right angles to EG; this will produce as sections (1) in the prism cut off from the whole prism a right-angled triangle, the base of which is MN, while the perpendicular is perpendicular at N to the plane $ABCD$ and equal to the axis of the cylinder, and the hypotenuse is in the plane cutting the cylinder, and (2) in the portion of the cylinder cut off a right-angled triangle the base of which is MO, while the perpendicular is the generator of the cylinder perpendicular at O to the plane KN, and the hypotenuse is.....................

[There is a lacuna here, to be supplied as follows.
Since $MN : NL = GK^2 : LS^2$
$= MN^2 : LS^2,$
it follows that $MN : ML = MN^2 : (MN^2 - LS^2)$
$= MN^2 : (MN^2 - MK^2)$
$= MN^2 : MO^2.$

But the triangle (1) in the prism is to the triangle (2) in the portion of the cylinder in the ratio of $MN^2 : MO^2$.

Therefore

($\triangle$ in prism) : ($\triangle$ in portion of cylinder)
$= MN : ML$
$=$ (straight line in rect. DG) : (straight line in parabola).

We now take all the corresponding elements in the prism, the portion of the cylinder, the rectangle DG and the parabola EFG respectively ;]

and it will follow that

(all the $\triangle$s in prism) : (all the $\triangle$s in portion of cylinder)
$=$ (all the str. lines in $\square\ DG$)
: (all the straight lines between parabola and EG).

But the prism is made up of the triangles in the prism, [the portion of the cylinder is made up of the triangles in it], the parallelogram DG of the straight lines in it parallel to KF, and the parabolic segment of the straight lines parallel to KF intercepted between its circumference and EG;

therefore (prism) : (portion of cylinder)
$= (\square\ GD)$: (parabolic segment EFG).

But $\square\ GD = \tfrac{3}{2}$ (parabolic segment EFG);
"for this is proved in my earlier treatise."

[*Quadrature of Parabola*]

Therefore prism $= \tfrac{3}{2}$ (portion of cylinder).

If then we denote the portion of the cylinder by 2, the prism is 3, and the original prism circumscribing the cylinder is 12 (being 4 times the other prism);

therefore the portion of the cylinder $= \tfrac{1}{6}$ (original prism).

Q.E.D.

[The above proposition and the next are peculiarly interesting for the fact that the parabola is an auxiliary curve introduced for the sole purpose of analytically reducing the required cubature to the known quadrature of the parabola.]

Proposition 14.

Let there be a right prism with square bases [and a cylinder inscribed therein having its base in the square $ABCD$ and touching its sides at E, F, G, H;
let the cylinder be cut by a plane through EG and the side corresponding to CD in the square face opposite to $ABCD$].

This plane cuts off from the prism a prism, and from the cylinder a portion of it.

It can be proved that the portion of the cylinder cut off by the plane is $\frac{1}{6}$ of the whole prism.

But we will first prove that it is possible to inscribe in the portion cut off from the cylinder, and to circumscribe about it, solid figures made up of prisms which have equal height and similar triangular bases, in such a way that the circumscribed figure exceeds the inscribed by less than any assigned magnitude..
....

But it was proved that
  (prism cut off by oblique plane)
  $< \frac{3}{2}$ (figure inscribed in portion of cylinder).
Now
  (prism cut off) : (inscribed figure)
    $= \square\ DG : (\square\text{s inscribed in parabolic segment})$;
therefore $\square\ DG < \frac{3}{2} (\square\text{s in parabolic segment})$:
which is impossible, since "it has been proved elsewhere" that the parallelogram DG is $\frac{3}{2}$ of the parabolic segment.

Consequently.........................
....not greater.
....

And (all the prisms in prism cut off)

: (all prisms in circumscr. figure)

= (all ▭s in ▭ DG)

: (all ▭s in fig. circumscr. about parabolic segmt.);

therefore

(prism cut off) : (figure circumscr. about portion of cylinder)

= (▭ DG) : (figure circumscr. about parabolic segment).

But the prism cut off by the oblique plane is > $\frac{3}{2}$ of the solid figure circumscribed about the portion of the cylinder..................... ...

....

[There are large gaps in the exposition of this geometrical proof, but the way in which the method of exhaustion was applied, and the parallelism between this and other applications of it, are clear. The first fragment shows that solid figures made up of prisms were circumscribed and inscribed to the portion of the cylinder. The parallel triangular faces of these prisms were perpendicular to GE in the figure of Prop. 13; they divided GE into equal portions of the requisite smallness; each section of the portion of the cylinder by such a plane was a triangular face common to an inscribed and a circumscribed right prism. The planes also produced prisms in the prism cut off by the same oblique plane as cuts off the portion of the cylinder and standing on GD as base.

The number of parts into which the parallel planes divided GE was made great enough to secure that the circumscribed figure exceeded the inscribed figure by less than a small assigned magnitude.

The second part of the proof began with the assumption that the portion of the cylinder is > $\frac{2}{3}$ of the prism cut off; and this was proved to be impossible, by means of the use of the auxiliary parabola and the proportion

$$MN : ML = MN^2 : MO^2$$

which are employed in Prop. 13.

We may supply the missing proof as follows*.

In the accompanying figure are represented (1) the first

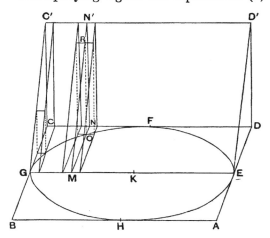

element-prism circumscribed to the portion of the cylinder, (2) two element-prisms adjacent to the ordinate OM, of which that on the left is circumscribed and that on the right (equal to the other) inscribed, (3) the corresponding element-prisms forming part of the prism cut off $(CC'GEDD')$ which is $\frac{1}{4}$ of the original prism.

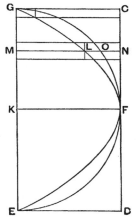

In the second figure are shown element-rectangles circumscribed and inscribed to the auxiliary parabola, which rectangles correspond exactly to the circumscribed and inscribed element-prisms represented in the first figure (the length of GM is the same in both figures, and the breadths of the element-rectangles are the same as the heights of the element-prisms);

---

* It is right to mention that this has already been done by Th. Reinach in his version of the treatise ("Un Traité de Géométrie inédit d'Archimède" in *Revue générale des sciences pures et appliquées*, 30 Nov. and 15 Dec. 1907); but I prefer my own statement of the proof.

the corresponding element-rectangles forming part of the rectangle GD are similarly shown.

For convenience we suppose that GE is divided into an even number of equal parts, so that GK contains an integral number of these parts.

For the sake of brevity we will call each of the two element-prisms of which OM is an edge "el. prism (O)" and each of the element-prisms of which MNN' is a common face "el. prism (N)." Similarly we will use the corresponding abbreviations "el. rect. (L)" and "el. rect. (N)" for the corresponding elements in relation to the auxiliary parabola as shown in the second figure.

Now it is easy to see that the figure made up of all the inscribed prisms is less than the figure made up of the circumscribed prisms by twice the final circumscribed prism adjacent to FK, i.e. by twice "el. prism (N)"; and, as the height of this prism may be made as small as we please by dividing GK into sufficiently small parts, it follows that inscribed and circumscribed solid figures made up of element-prisms can be drawn differing by less than any assigned solid figure.

(1) Suppose, if possible, that

(portion of cylinder) $> \frac{2}{3}$ (prism cut off),

or (prism cut off) $< \frac{3}{2}$ (portion of cylinder).

Let (prism cut off) $= \frac{3}{2}$(portion of cylinder $- X$), say.

Construct circumscribed and inscribed figures made up of element-prisms, such that

(circumscr. fig.) $-$ (inscr. fig.) $< X$.

Therefore (inscr. fig.) $>$ (circumscr. fig. $- X$),
and *a fortiori* $>$ (portion of cyl. $- X$).

It follows that

(prism cut off) $< \frac{3}{2}$ (inscribed figure).

Considering now the element-prisms in the prism cut off and those in the inscribed figure respectively, we have

el. prism (N) : el. prism $(O) = MN^2 : MO^2$
$= MN : ML$ [as in Prop. 13]
$=$ el. rect. (N) : el. rect. (L).

It follows that

Σ {el. prism (N)} : Σ {el. prism (O)}
= Σ {el. rect. (N)} : Σ {el. rect. (L)}.

(There are really two more prisms and rectangles in the first and third than there are in the second and fourth terms respectively; but this makes no difference because the first and third terms may be multiplied by a common factor as $n/(n-2)$ without affecting the truth of the proportion. Cf. the proposition from *On Conoids and Spheroids* quoted on p. 15 above.)

Therefore

(prism cut off) : (figure inscr. in portion of cyl.)
= (rect. GD) : (fig. inscr. in parabola).

But it was proved above that

(prism cut off) < $\frac{3}{2}$ (fig. inscr. in portion of cyl.);

therefore (rect. GD) < $\frac{3}{2}$ (fig. inscr. in parabola),

and, *a fortiori*,

(rect. GD) < $\frac{3}{2}$ (parabolic segmt.):

which is impossible, since

(rect. GD) = $\frac{3}{2}$ (parabolic segmt.).

Therefore

(portion of cyl.) is *not* greater than $\frac{2}{3}$ (prism cut off).

(2) In the second lacuna must have come the beginning of the next *reductio ad absurdum* demolishing the other possible assumption that the portion of the cylinder is < $\frac{2}{3}$ of the prism cut off.

In this case our assumption is that

(prism cut off) > $\frac{3}{2}$ (portion of cylinder);

and we circumscribe and inscribe figures made up of element-prisms, such that

(prism cut off) > $\frac{3}{2}$ (fig. circumscr. about portion of cyl.).

We now consider the element-prisms in the prism cut off and in the circumscribed figure respectively, and the same argument as above gives

(prism cut off) : (fig. circumscr. about portion of cyl.)
$= $ (rect. GD) : (fig. circumscr. about parabola),

whence it follows that

(rect. GD) $> \frac{3}{2}$ (fig. circumscribed about parabola),

and, *a fortiori*,

(rect. GD) $> \frac{3}{2}$ (parabolic segment):

which is impossible, since

(rect. GD) $= \frac{3}{2}$ (parabolic segmt.).

Therefore

(portion of cyl.) is *not* less than $\frac{2}{3}$ (prism cut off).

But it was also proved that neither is it greater; therefore (portion of cyl.) $= \frac{2}{3}$ (prism cut off)
$= \frac{1}{6}$ (original prism).]

[**Proposition 15.**]

[This proposition, which is lost, would be the mechanical investigation of the second of the two special problems mentioned in the preface to the treatise, namely that of the cubature of the figure included between two cylinders, each of which is inscribed in one and the same cube so that its opposite bases are in two opposite faces of the cube and its surface touches the other four faces.

Zeuthen has shown how the mechanical method can be applied to this case*.

In the accompanying figure $VWYX$ is a section of the cube by a plane (that of the paper) passing through the axis BD of one of the cylinders inscribed in the cube and parallel to two opposite faces.

The same plane gives the circle $ABCD$ as the section of the other inscribed cylinder with axis perpendicular to the

---

* Zeuthen in *Bibliotheca Mathematica* VII_3, 1906–7, pp. 356–7.

plane of the paper and extending on each side of the plane to a distance equal to the radius of the circle or half the side of the cube.

AC is the diameter of the circle which is perpendicular to BD.

Join AB, AD and produce them to meet the tangent at C to the circle in E, F.

Then $EC = CF = CA$.

Let LG be the tangent at A, and complete the rectangle $EFGL$.

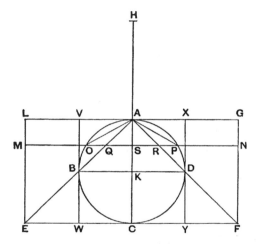

Draw straight lines from A to the four corners of the section in which the plane through BD perpendicular to AK cuts the cube. These straight lines, if produced, will meet the plane of the face of the cube opposite to A in four points forming the four corners of a square in that plane with sides equal to EF or double of the side of the cube, and we thus have a pyramid with A for vertex and the latter square for base.

Complete the prism (parallelepiped) with the same base and height as the pyramid.

Draw in the parallelogram LF any straight line MN parallel to EF, and through MN draw a plane at right angles to AC.

This plane cuts—

(1) the solid included by the two cylinders in a square with side equal to OP,
(2) the prism in a square with side equal to MN, and
(3) the pyramid in a square with side equal to QR.

Produce CA to H, making HA equal to AC, and imagine HC to be the bar of a balance.

Now, as in Prop. 2, since $MS = AC$, $QS = AS$,
$$MS \cdot SQ = CA \cdot AS$$
$$= AO^2$$
$$= OS^2 + SQ^2.$$
Also
$$HA : AS = CA : AS$$
$$= MS : SQ$$
$$= MS^2 : MS \cdot SQ$$
$$= MS^2 : (OS^2 + SQ^2), \text{ from above,}$$
$$= MN^2 : (OP^2 + QR^2)$$
$$= (\text{square, side } MN) : (\text{sq., side } OP + \text{sq., side } QR).$$

Therefore the square with side equal to MN, in the place where it is, is in equilibrium about A with the squares with sides equal to OP, QR respectively placed with their centres of gravity at H.

Proceeding in the same way with the square sections produced by other planes perpendicular to AC, we finally prove that the prism, in the place where it is, is in equilibrium about A with the solid included by the two cylinders and the pyramid, both placed with their centres of gravity at H.

Now the centre of gravity of the prism is at K.

Therefore $HA : AK = (\text{prism}) : (\text{solid} + \text{pyramid})$
or $\qquad 2 : 1 = (\text{prism}) : (\text{solid} + \frac{1}{3} \text{ prism})$.

Therefore $2 (\text{solid}) + \frac{2}{3} (\text{prism}) = (\text{prism})$.

It follows that
$$(\text{solid included by cylinders}) = \tfrac{1}{6} (\text{prism})$$
$$= \tfrac{2}{3} (\text{cube}). \quad \text{Q.E.D.}$$

THE METHOD

There is no doubt that Archimedes proceeded to, and completed, the rigorous geometrical proof by the method of exhaustion.

As observed by Prof. C. Juel (Zeuthen *l.c.*), the solid in the present proposition is made up of 8 pieces of cylinders of the type of that treated in the preceding proposition. As however the two propositions are separately stated, there is no doubt that Archimedes' proofs of them were distinct.

In this case AC would be divided into a very large number of equal parts and planes would be drawn through the points of division perpendicular to AC. These planes cut the solid, and also the cube VY, in square sections. Thus we can inscribe and circumscribe to the solid the requisite solid figures made up of element-prisms and differing by less than any assigned solid magnitude; the prisms have square bases and their heights are the small segments of AC. The element-prism in the inscribed and circumscribed figures which has the square equal to OP^2 for base corresponds to an element-prism in the cube which has for base a square with side equal to that of the cube; and as the ratio of the element-prisms is the ratio $OS^2 : BK^2$, we can use the same auxiliary parabola, and work out the proof in exactly the same way, as in Prop. 14.]